Advances in Sustainability Science and Technology

The book series aims at bringing together valuable and novel scientific contributions that address the critical issues of renewable energy, sustainable building, sustainable manufacturing, and other sustainability science and technology topics that have an impact in this diverse and fast-changing research community in academia and industry.

The areas to be covered are

- Climate change and mitigation, atmospheric carbon reduction, global warming
- Sustainability science, sustainability technologies
- Sustainable building technologies
- Intelligent buildings
- Sustainable energy generation
- Combined heat and power and district heating systems
- Control and optimization of renewable energy systems
- Smart grids and micro grids, local energy markets
- Smart cities, smart buildings, smart districts, smart countryside
- Energy and environmental assessment in buildings and cities
- Sustainable design, innovation and services
- Sustainable manufacturing processes and technology
- Sustainable manufacturing systems and enterprises
- Decision support for sustainability
- Micro/nanomachining, microelectromechanical machines (MEMS)
- Sustainable transport, smart vehicles and smart roads
- Information technology and artificial intelligence applied to sustainability
- Big data and data analytics applied to sustainability
- Sustainable food production, sustainable horticulture and agriculture
- Sustainability of air, water and other natural resources
- Sustainability policy, shaping the future, the triple bottom line, the circular economy

High quality content is an essential feature for all book proposals accepted for the series. It is expected that editors of all accepted volumes will ensure that contributions are subjected to an appropriate level of reviewing process and adhere to KES quality principles.

The series will include monographs, edited volumes, and selected proceedings.

More information about this series at http://www.springer.com/series/16477

Binod Kumar Kanaujia · Neeta Singh ·
Sachin Kumar

Rectenna: Wireless Energy Harvesting System

Springer

Binod Kumar Kanaujia
School of Computational and Integrative Sciences
Jawaharlal Nehru University
New Delhi, India

Neeta Singh
School of Computational and Integrative Sciences
Jawaharlal Nehru University
New Delhi, India

Sachin Kumar
Department of Electronics
and Communication Engineering
SRM Institute of Science and Technology
Chennai, India

ISSN 2662-6829 ISSN 2662-6837 (electronic)
Advances in Sustainability Science and Technology
ISBN 978-981-16-2535-0 ISBN 978-981-16-2536-7 (eBook)
https://doi.org/10.1007/978-981-16-2536-7

This Springer imprint is published by the registered company Springer Nature Singapore Pte Ltd.
The registered company address is: 152 Beach Road, #21-01/04 Gateway East, Singapore 189721, Singapore

Preface

Wireless energy harvesting is in high demand these days due to its numerous applications. Understanding energy harvesting methods is essential for real-time applications. The goal of this book is to give readers an overview and approach to designing harvesting systems. Electromagnetic analysis and a literature review of various harvesting systems are also included in this book.

Our book is divided into five chapters.

The first chapter, "Introduction," introduces wireless energy harvesting and the device used in this process, known as a "rectenna." This chapter discusses various types of energy sources that are freely available in the environment. Previously published work and recent developments are also discussed to provide a better understanding of the harvesting system.

The second chapter "Background and Origin of the Rectenna" presents a history of rectenna. It discusses the architecture of the harvesting system. It provides a detailed overview and development of the rectenna. The impact of various energy harvesting applications on human life is also explained.

The third chapter "Antennas" investigates various types of antennas used in energy harvesting systems. The benefits and drawbacks of each antenna type are discussed. The most recent advancements in antenna design are also explained.

The fourth chapter, "Matching Network and Rectifier Circuit," discusses the importance of the rectifier and impedance matching circuit. This chapter discusses various lumped circuits and distributed circuits, as well as their advantages and drawbacks. The theory and various topologies of rectifier and matching circuits are discussed in depth. The characteristics of the Schottky diode are also discussed.

The fifth chapter, "Rectenna Implementation," provides an overview of various techniques available for designing a rectenna. Various electromagnetic simulators are discussed for antenna and rectifier circuit implementation. Optimization techniques are also discussed to help readers understand the antenna and rectifier integration.

New Delhi, India — Binod Kumar Kanaujia
New Delhi, India — Neeta Singh
Chennai, India — Sachin Kumar

Contents

About the Authors

Binod Kumar Kanaujia received the B.Tech. degree from Kamla Nehru Institute of Technology, Sultanpur, India, in 1994, and the M.Tech. and Ph.D. degrees from the Indian Institute of Technology Banaras Hindu University, Varanasi, India, in 1998 and 2004, respectively. He is currently a Professor at the School of Computational and Integrative Sciences, Jawaharlal Nehru University, New Delhi, India. He has been credited to publish more than 300 research papers in several peer-reviewed journals and conferences. He has supervised 50 M.Tech. and 15 Ph.D. scholars in the field of RF and microwave engineering. He is currently on the editorial board of several international journals. He is a member of the Institution of Engineers, India, the Indian Society for Technical Education, and the Institute of Electronics and Telecommunication Engineers of India. He had successfully executed five research projects sponsored by different agencies of the Government of India, i.e., DRDO, DST, AICTE, and ISRO.

Neeta Singh received the B.Tech. and M.Tech. degrees from Guru Gobind Singh Indraprastha University, Delhi, India, in 2012 and 2015, respectively, and the Ph.D. degree from Jamia Millia Islamia, Delhi, India, in 2020. She is currently a Research Fellow at the Jawaharlal Nehru University, New Delhi, India. She is a recipient of the Teaching-cum-Research Fellowship from the Government of NCT of Delhi, India. Her current research interest includes microstrip antennas, rectenna, and green energy technology.

Sachin Kumar received the B.Tech. degree from Uttar Pradesh Technical University, Lucknow, India, in 2009, and the M.Tech. and Ph.D. degrees from Guru Gobind Singh Indraprastha University, Delhi, India, in 2011 and 2016, respectively. From 2018 to 2020, he was a Post-Doctoral Fellow at the College of IT Engineering, Kyungpook National University, South Korea. He is currently a Research Assistant Professor at the SRM Institute of Science and Technology, Chennai, India. He has published over a hundred research articles in several peer-reviewed international journals and conferences. He serves as the session chair, organizer, and member of the program committee for various conferences, workshops, and short courses in electronics and computer-related topics. He is also a frequent reviewer for more than fifty scientific journals and book publishers. He is a recipient of the Teaching-cum-Research Fellowship from the Government of NCT of Delhi, India, and the Brain Korea 21 Plus Research Fellowship from the National Research Foundation of South Korea. He is a member of the Indian Society for Technical Education and the Korean Institute of Electromagnetic Engineering and Science.

Chapter 1
Introduction

1.1 Background

"Rectenna" is a device used to energize the low-power systems without using any wired connections [1, 2]. The rectenna or rectifying antenna is also called a wireless battery. Have you ever heard about the concept of wireless electricity? Yes, a rectenna can provide wireless electricity to the home appliances, medical equipment's in the healthcare centers, electronic systems in the school/colleges, etc. A rectenna mainly consists of a sensing antenna and a rectifier. The receiving or sensing antenna is the main component that collects electromagnetic (EM) signals present in the nearby surroundings [3, 4]. The collected energy is the waste energy produced from sources such as cell-phone towers, laptops, mobile phones, satellite systems, TV towers, and Wi-Fi routers [5]. With the development of wire-free devices/systems, various researches have been carried out to provide wireless electricity to the low-power electrical/electronic devices. Recently, wireless power transmission (WPT) technology has received a lot of importance due to its clean and green nature [6, 7].

1.2 Why Not Wires?

Our future will be based on wireless technology, where we would be able to access electricity with the help of wireless power hotspots. N. Tesla, the famous scientist, was the first person to perform an experiment on WPT in 1899 [8]. But, what is wrong with the wires?

Wired power transmission has multiple disadvantages as compared to the WPT. Some of them are listed below:

- In wired power transmission and distribution, the losses are ~26–40%.
- More chances of getting electric shock to humans and animals.

B. K. Kanaujia et al., *Rectenna: Wireless Energy Harvesting System*,
Advances in Sustainability Science and Technology,
https://doi.org/10.1007/978-981-16-2536-7_1

- Health and environmental hazards.

As technology is advancing day by day, an increased number of electronic devices are being used in our daily life. These devices have a limited lifeline, and many of them can be used once only [9, 10].

1.2.1 E-waste

The electronic-waste consists of various materials like copper, gold, lithium, platinum, silver, etc. These materials contain harmful and hazardous chemicals, such as arsenic, brominate flame retardants, cadmium, lead, mercury, polychlorinated, and PVC plastic, as shown in Fig. 1.1.

The impact of e-waste materials on the body is explained below:

- Lead: It damages the brain and may also cause coma when present in a large amount. It can cause reading and learning disabilities and reduce IQ level and sometimes hearing loss.
- Arsenic: It can damage the nervous system and can cause skin diseases. It increases the risk of cancer in the lungs, kidneys, and liver.
- Cadmium: A long time contact with cadmium weakens the bones and can cause kidney damage.
- Mercury: It is hazardous for health and the environment. It directly attacks the central nervous system and the immune system. It also acts as a poison for the kidneys, eyes, and skin.

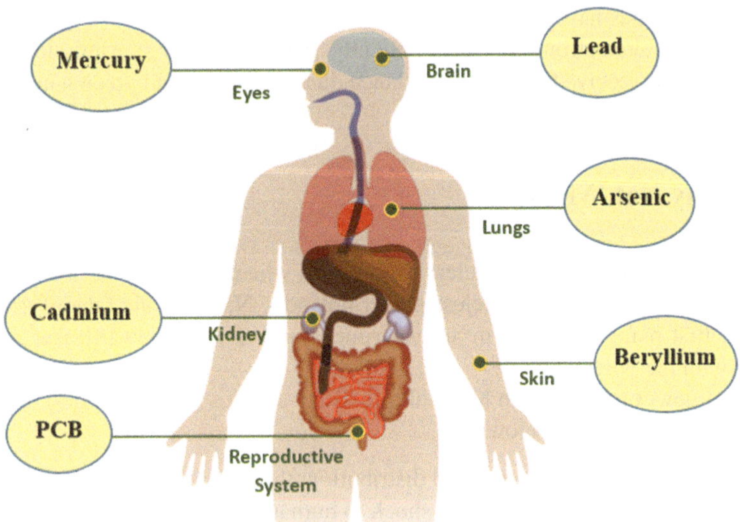

Fig. 1.1 Effect of e-waste on the human body

- Polychlorinated biphenyl (PCB): It damages the immune system and reproductive system of both males and females. It is used in the manufacturing of transistors and capacitors.

1.2.2 Power Loss

A large transmission loss is seen during the transfer of electrical energy through wires. Long distance between wires and transmission lines dissipates a large amount of power in the electrical energy system.

$$\eta = \frac{\text{Power sent} - \text{Power loss in the line}}{1 - \text{Power loss in the line}} \tag{1.1}$$

$$\text{Tranmitted Power}(P) = \sqrt{3} \times VI\cos\theta \tag{1.2}$$

where $\cos\theta$ is the power factor, V is the voltage, and I is the current.

$$\text{Load Factor} = \frac{\text{Average load in a specific time} - \text{period}}{\text{peak load during that time} - \text{period}} \tag{1.3}$$

The electrical power loss can be defined as

$$P_{\text{loss}} = 3R(Tc) \times I^2 \times L \tag{1.4}$$

where I is the line current, L is the line length, and R is the resistance of the wire.

The main factors for power loss are:

- In rural areas, the distribution wires are spread over a long distance, which causes high resistance loss and a large power drop.
- For lossless transmission lines, the power factor should be very high. But, in some areas, it is around 0.65–0.75, which causes distribution losses.
- Short size of conductors used in the transmission lines.
- The load factor is also responsible for losses.
- A higher number of joints produce high power loss.
- 16.6% of power losses are due to error in meter reading, defective meter, etc.

1.2.2.1 Electric Shock Hazards

The factors responsible for electric shock are:

- When the body comes in contact with the cables.
- Due to broken wires or cable faults.
- In an explosive environment, the wire can work as a source for fire or blast.
- This may also occur due to an ungrounded electrical system.

1.3 Wireless Power Transmission (WPT)

As the name suggests, WPT transfers power without using any wires or cables. With this technology, power can be easily transmitted from one destination to another destination [11, 12]. The different types of WPT technology may include energy sources like magnetic energy, microwave energy, laser energy, and solar energy [13–16]. The working of wireless power transfer can be understood by the following steps:

- Let us consider two identical antennas or coil, which resonates at the same frequency, named as coil-A/antenna-A and coil-B/antenna-B. The magnetic coil-A is fitted inside a box and is kept at any ceiling or wall.
- Antenna-A receives power from the main power supply, and it is connected through a cord.
- The EM waves generated from the antenna-A/coil-A travel through the air.
- Antenna-B/coil-B, acting as a receiver coil, may be embedded into any electronic gadget or home appliances. It resonates at the frequency of the transmitting antenna and collects EM energy from it.

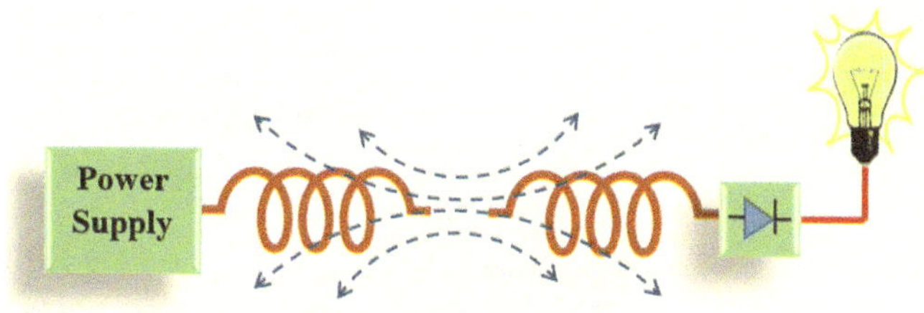

Fig. 1.2 Schematic of the WPT system

- The collected EM energy is converted into DC power using a rectifier.

It works on the basic rule of physics as given below:

- At source: The varying electric field provides a varying magnetic field.
- At destination: The changing magnetic field provides electricity that charges the device as shown in Fig. 1.2.

1.3.1 Need of WPT

In today's world, it is difficult to imagine life without electricity. The wires and batteries are frequently used to charge or transmit power supply without thinking about their harmful effects. Wires are made of copper and aluminum metals, which are costly and their decomposition is harmful to the environment. Figure 1.3 shows the importance and advantages of WPT.

1.3.1.1 Advantages of WPT

- No wire, no e-waste: Every machine is now connected with the wires and most of the wires have a plastic covering, which is made of dangerous chemicals. This plastic is extremely hazardous for the earth if discarded recklessly into the landfills as highlighted in Fig. 1.4. The decomposition of e-waste is becoming a serious issue globally.
- Transmission of power to remote areas: WPT can supply power to the remote area, such as mountains, villages, forests, red zone areas, where the use of wires may be hazardous or installation of any wired system seems to be impractical. However, the urban areas are also moving toward the wireless power supply [17] as wireless Internet (Wi-Fi) zones are being set up on the large scale. Similarly, in the coming years, the Wi-power zones need to be commercialized at various places such as restaurants, streets, and shopping malls [18].
- Wireless power grid: The electricity can be made available to the rural population through the WPT technique [19, 20]. The conventional infrastructure or grid

Fig. 1.3 Advantages of WPT

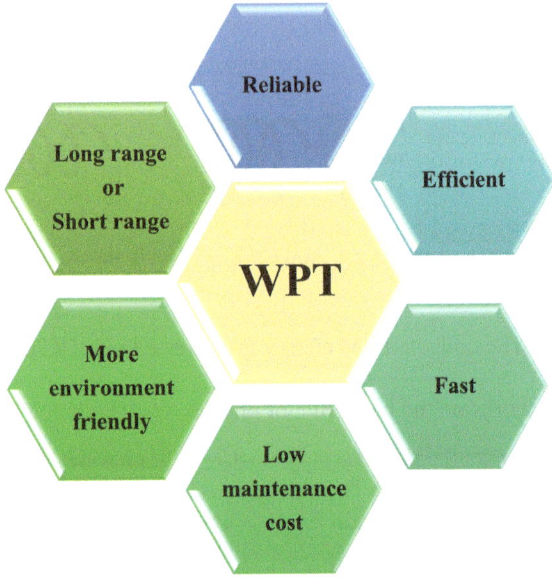

extension can be replaced by the WPT grid, which will reduce the cost of infrastructure used in the distribution of electricity (Fig. 1.5). The performance of the conventional and wireless grid is compared in Table 1.1.

- Line of sight (LoS) shows advantages in WPT: It is the transmission of a signal through a straight line between the transmitter and the receiver. For the short-range, LoS transmission is preferred to avoid the beam spreading. If the LoS does not exist, the power will be reflected, refracted, or scattered in the air resulting in loss of energy.

Fig. 1.4 Landfills covered with e-waste

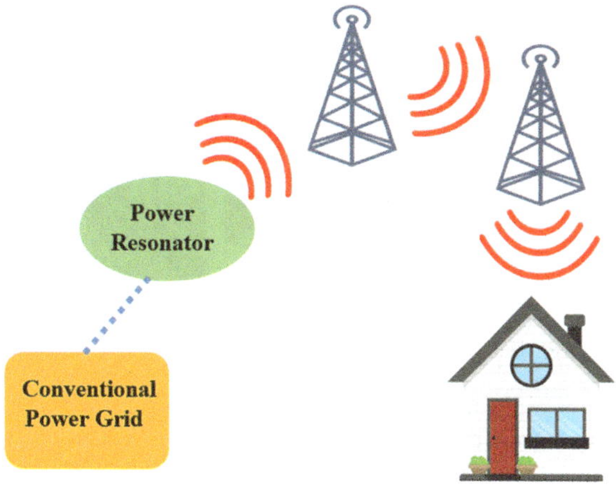

Fig. 1.5 Wireless power grid system

Table 1.1 Comparison of conventional and wireless grid

Infrastructure type	Conventional grid	Wireless grid
Transmission mode	Wired network, electric pole	No wires, power repeaters
Distribution	Distribution poles, congested wired lines	Less congestion, wireless distribution
Economic assessment	High cost, environmental damage	Eco-friendly, clean, and green technology

1.4 Wireless Energy Harvesting (WEH)

Energy harvesting is the collection of ambient energy present in the nearby surroundings. The harvested energy can be directly used to energize the low-power electronic devices or recharge the secondary devices [21–23]. The use of electronic systems is increasing day by day for home automation, IoT applications, industrial applications, smart cities, etc. The use of wires and batteries leads to environmental pollution, shown in Fig. 1.6. The battery is an important source of power for electronic devices; however, it shows hazardous effects, like the decomposition of materials used to manufacture it are harmful and its short life is a serious issue [24, 25]. As many systems depend on these batteries and even cannot afford the replacement, the cost of these devices is also a concern.

The ambient energies include solar energy [26], wind energy [27], vibrational energy [28], thermal energy [29], and RF energy [30]. Out of these available energy sources, RF energy has better availability as it does not depend on the climate, and it is cost-effective as the components used for the conversion are very cheap. Also, the setup used for RF energy harvesting is compact size, unlike a wind turbine setup

Fig. 1.6 Batteries and wires end up polluting the environment

that requires a large area for installation. The different types of energy sources are compared in Table 1.2. A detailed investigation of different available energy sources is as follows:

- **Solar energy**: A photovoltaic cell (PVC) is used to convert solar radiation or artificial light into electrical power. The PVC consists of two different materials doped together to form a p–n junction. It generates high output power ranging from microwatts to milliwatts based on various factors like cell size, the intensity of light, and the position of the cell elements to obtain the maximum conversion efficiency. The environmental conditions and the availability of light may also limit the performance of the PVC.
- **Wind energy**: The wind energy is converted into electrical energy through rotors and turbines. The conversion efficiency depends on the wind strength and its flow rate. A mini-wind turbine can also be used to harvest the wind energy; however, it shows low power as compared to a large-scale wind turbine.
- **Vibrational energy**: Any vibration or mechanical movement can be converted into electrical energy. Various techniques are available to harvest vibrational energy, but the piezoelectric method is the most commonly used.
  The piezoelectric material contains a crystalline structure, and, here, the negative and positive charges do not overlap. But, the electrical charge is generated due to the fluctuations in the dipole moment on account of the mechanical strain on the

Table 1.2 Comparison of different energy sources

Energy sources	Availability	Power density	Setup size	Advantages
Solar energy	Only in day time	100 mW/cm^2	Large	Limitless availability
Wind energy	Depends on climate	3.5 mW/cm^2	Large	High power density
Thermal energy	Regularly	60 $\mu W/cm^2$	Average	Climate independent
Vibrational energy	Activity dependent	200 $\mu W/cm^3$	Average	Both indoor and outdoor applications
RF energy	Continuous	1 $\mu W/cm^2$	Small	Anytime, anywhere present

piezoelectric material. This material harvests the vibrational energy and converts it into DC power.

- **Thermal energy**: Thermal energy can be converted by using three different methods: Thomson effect, Seebeck effect, and Peltier effect. The converted energy is mainly used for outdoor applications, such as the heat energy from roadway pavement can be captured and converted into electrical energy by the generators embedded near the road.
- **RF energy**: Unlike other energy sources, RF energy is present both day and night. The two types of RF energy sources are:

Dedicated energy sources: These sources depend on demand–supply and use a license-free band to design the harvesting system. As it is a dedicated source, it is very directive toward the receiver, and high-power density can be achieved from it, which increases the conversion efficiency [31].

Ambient energy sources: These sources are present freely in the environment, but they provide less amount of power or wireless electricity to charge the batteries as compared to the dedicated sources. The amount of harvested energy depends on various parameters like transmitted power, received power, and the RF wavelength [32]. All these parameters can be calculated using the Friss transmission equation [33].

1.4.1 Friss Transmission Equation

Friss transmission equation is the key source for wireless communication. It calculates the power received at the receiving antenna.

$$P_D = \frac{P_\mathrm{t}}{4\pi r^2} \tag{1.5}$$

where P_t is the transmitting power and r is the distance between the transmitter and the receiver.

The above case demonstrates when transmitting antenna generates power isotopically and the directivity is 0 dBm. If the directivity is more than 0 dBm, the gain of the antenna (G_t) is also considered.

$$P_D = \frac{P_\mathrm{t}}{4\pi r^2} G_t \tag{1.6}$$

The power received at the receiving antenna considering the effective area of the antenna is

$$P_r = P_D * A_\mathrm{e} \tag{1.7}$$

where A_e is the effective aperture area, which is defined as the area on which power is incident and captured.

$$A_e = \frac{G_r.\lambda^2}{4\pi} \tag{1.8}$$

By combining Eqs. (1.6) and (1.8), Eq. (1.7) becomes

$$P_r = \frac{P_t G_t}{4\pi r^2} \frac{G_r.\lambda^2}{4\pi} = \frac{P_t G_t G_r \lambda^2}{(4\pi r)^2} \tag{1.9}$$

The factors affecting the power transmission are:

- **Antenna direction**: The receiving and transmitting antennas should be pointed to each other to obtain the maximum gain. The antenna with high gain shows the high conversion efficiency.
- **Impedance of the circuit**: The antenna impedance should be perfectly matched with the load impedance to achieve maximum power transfer (MPT). According to the MPT theorem, the load impedance is perfectly matched with the conjugate of source impedance; hence, the complex part cancels out and the real impedance part is left ($Z_L = Z_S$*).
- **Lossless polarization**: Polarization is a big factor in wireless communication. It is the orientation of the wave traveling in the direction of propagation. If the transmitting polarization mismatches with the receiver then, the received wave has a high amount of distortion, which results in low conversion efficiency.

1.5 Frequency Range of the Rectenna

The RF energy present in the form of EM waves is measured in terms of frequency or wavelength. How much accessible RF energy is available in the environment? How it can be used for some needful purpose? Is it harmful if we remain surrounded by it for a longer time? These questions come to our mind when we learn about the RF/EM signal.

The EM sources are not good for human beings and animals [34, 35]. The wasted RF energy can be collected and converted into DC to power up the low-power electronic devices or to recharge the batteries. First, let us understand what exactly the EM waves are? EM wave is a combination of changing magnetic and electric fields radiated by a source. The EM wave propagates in a direction perpendicular to the electric and magnetic fields, which means all three fields oscillate in a perpendicular phase. The famous physicist J.C. Maxwell stated that the propagating EM waves have some velocity in free space, which is defined as

$$\upsilon = \frac{1}{\sqrt{\mu_0 \varepsilon_0}} \tag{1.10}$$

where μ_0 is the permeability in free space and ε_0 is the permittivity in free space.

The relative permittivity and absolute permeability are also considered when the wave propagates in a medium, which are evaluated as

$$\mu = \mu_0 \mu_r \tag{1.11}$$

$$\varepsilon = \varepsilon_0 \varepsilon_r \tag{1.12}$$

where ε_r is 1 for the free-space medium, and it can vary between 1 and 10 according to the medium. μ_r is also 1 for free-space medium and for non-magnetic material. In the microwave range, mostly the non-magnetic materials are used.

$$\upsilon = \frac{1}{\sqrt{\mu_0 \varepsilon_0 \varepsilon_r}} \tag{1.13}$$

$$\upsilon = \frac{c}{\sqrt{\varepsilon_r}}, \quad \text{where } c = \frac{1}{\sqrt{\mu_0 \varepsilon_0}} \tag{1.14}$$

The characteristics of EM waves can be defined with the help of Maxwell's equations, for example, how they propagate in the material or how they interact with the material properties and show its effects.

- **Ampere's law**

It states that a line integral for any closed-loop path of magnetic flux is equal to the total current passing through the close loop.

$$\oint \vec{H}.dl = I \tag{1.15}$$

J_T is the total current density defined as

$$J_T = J + \frac{\partial D}{\partial t} \tag{1.16}$$

Here, J_T is changed with I as

$$\oint \vec{H}.dl = \oint_s \left(J + \frac{\partial D}{\partial t} \right).ds \tag{1.17}$$

Applying Stoke's theorem, this states that the line integral of a vector field around the surface boundary is equal to the surface integral of the curl of that vector.

$$\oint \vec{H}.dl = (\nabla \times H).ds \tag{1.18}$$

When comparing Eqs. (1.17) and (1.18)

$$\oint (\nabla \times H).ds = \oint_{s} \left(J + \frac{\partial D}{\partial t} \right).ds \tag{1.19}$$

$$(\nabla \times H) = J + \frac{\partial D}{\partial t} \tag{1.20}$$

where I is the total enclosed current, dl is the closed-loop incremental loop, J is the conduction current density, and $\frac{\partial D}{\partial t}$ is the displacement current density.

- **Faraday's law**

It states that if there is a change in magnetic flux, the electromotive force (EMF) is induced in the circuit.

$$V = -\frac{\partial \Phi}{\partial t} \tag{1.21}$$

$$\Phi = \oint_{s} B.ds \text{ and } V = \oint_{s} E.dl \tag{1.22}$$

On comparing and putting the values of Eq. (1.22) in (1.21), it becomes

$$\oint_{s} E.dl = -\frac{\partial}{\partial t} \oint_{s} B.ds \tag{1.23}$$

By applying the Stoke's theorem defined in Eq. (1.18)

$$\oint_{s} (\nabla \times E).ds = -\frac{\partial}{\partial t} \oint_{s} B.ds \tag{1.24}$$

$$\nabla \times E = -\frac{\partial B}{\partial t} \tag{1.25}$$

- **Gauss's law**

It states that the electric flux enclosed in a closed surface is directly proportional to the total charge enclosed per unit volume.

$$\oint_{s} D.ds = \oint_{v} \rho.dv \tag{1.26}$$

Applying divergence theorem

$$\lim_{dV \to 0} \int \frac{D.ds}{dV} = \nabla.D \tag{1.27}$$

Putting Eq. (1.27) into (1.28), the Gauss's equation is

$$\nabla.D = \rho \tag{1.28}$$

The summary of Maxwell's equation is shown in Table 1.3.

The energy harvesting focuses on the freely available energy present in the microwave range from 500 MHz to 2.5 GHz. The microwave term defines the EM waves propagating between 0.3 GHz and 30 GHz with a wavelength range from 1 cm to 1 m. EM waves beyond the microwave range, from 30 to 300 GHz, are termed as millimeter waves. Infrared (IR) has a range from 300 GHz to 400 THz [36, 37]. Beyond IR, there is an optical spectrum, which is also called visible light, ranges from 400 to 770 THz. Further, ultraviolet (UV) rays (from 750×10^{12} to 30×10^{15} Hz) and gamma rays are present. The microwave frequency band is further sub-divided into L, S, C, X, Ku, K, and Ka bands. The EM spectrum is shown in Table 1.4.

Microwave has various applications, like point-to-point communication, where information is transmitted in the form of data, video, and audio. The second important use of microwave is for radio detection and ranging (radar) applications [38], where the location of the target is found. It can be achieved using a directional antenna and is based on the principle of Doppler effect. When the target approaches toward the source, the returning microwave signals are more compressed as compared to the transmitted microwave signals and have a high frequency or short wavelength. Similarly, if the target is going away from the source, then the returning microwave signal has a higher wavelength or low frequency. This frequency shift describes the speed of the target [39]. In the microwave range, other applications of the radar are collision avoidance, SONAR, missile guidance, air traffic control, and speed limit enforcement. The microwave heat source is also very common and used ordinarily. The microwave oven is the most popular application of microwaves and is used

Table 1.3 Summary of Maxwell's equation

Name	Integral form	Differential form	Remarks
Gauss's law (electric)	$\oint \vec{E}.d\vec{A} = \frac{Q_{enc}}{\varepsilon_0}$	$\nabla.\vec{E} = \frac{\rho}{\varepsilon_0}$	Relationship between electric field and electric charge
Gauss's law (magnetic)	$\oint \vec{B}.d\vec{A} = 0$	$\nabla.\vec{B} = 0$	Net magnetic flux out of any closed surface is zero
Faraday's law of induction	$\oint \vec{E}.d\vec{s} = -\frac{d\Phi_B}{dt}$	$\nabla \times \vec{E} = -\frac{\partial \vec{B}}{\partial t}$	Charging magnetic field produces charging electric field
Ampere's circuital law	$\oint \vec{B}.d\vec{s} = -\frac{d\Phi_E}{dt}$	$\nabla \times \vec{H} = \vec{J} + \frac{\partial \vec{D}}{\partial t}$	Time varying electric field produces magnetic field

Table 1.4 EM spectrum

Wavelength	Frequency	Designation
10^4–10^3 km	30–300 Hz	Extremely low frequency (ELF)
10^3–100 km	300–3000 Hz	Ultra-low frequency (ULF)
100–10 km	3–30 kHz	Very low frequency (VLF)
10–1 km	30–300 kHz	Low frequency (LF)
1 km–100 m	300–3000 kHz	Medium frequency (MF)
100–10 m	3–30 MHz	High frequency (HF)
10–1 m	30–300 MHz	Very high frequency (VHF)
1–0.1 m	3–3000 MHz	Ultra-high frequency (UHF)
0.1–0.01 m	3–30 GHz	Super-high frequency (SHF)
	30–300 GHz	Extremely high frequency (ELF)

for industrial purpose, chemical industry, cooking food, and in the field of health care [40–42]. Another application of microwaves is satellite communication, where satellites are used as a relay station. It uses two different frequencies range for uplink (5.9–6.4 GHz) and downlink (3.7–4.2 GHz). The satellites are used for military, civilian, and surveillance applications. The microwaves are preferred due to the following reasons:

- The microwave frequency range is very high, which shows a wider bandwidth. The problem to accommodate more users or more number of applications can be solved by increasing the transmission bandwidth range.
- The antenna size is inversely proportional to the frequency; therefore, the antenna size decreases with the frequency, and thus, it can be easily integrated into any wireless device like aircraft, drones, missiles, mobile phones, etc.
- Microwave signals are generally used for satellite communication due to its transparency property, which defines that it can propagate in the atmosphere with very less attenuation as compared to other frequencies like visible rays, x-rays, etc. Therefore, satellite communication is easily possible with microwaves as it is capable of penetrating through the ionosphere with a very little reflection and absorption.
- An important property of microwave frequency is the LoS. It reduces the fading effect; thus, the data or signal can be received with more intensity. Directional beam antennas help in long-distance communication with very less power consumption.
- Wireless communication is one of the common uses of microwave frequency as it can pass through obstacles like buildings, doors, hills, walls, etc.

The 2.45 GHz frequency is frequently used by the telecommunication industry and electronic gadgets. Some of the advantages of 2.45 GHz are described below:

- This band is free to use. It is assigned as industrial scientific and medical (ISM) band, which can be used without any license.
- While designing any electronics or wireless gadgets, the makers want to ensure reliable and lossless data transfer. The ISM band does not interfere with the AM radio band (535 kHz–1.7 MHz), FM broadcasting band (88–108 MHz), and GSM band (0.8/1.9 GHz).
- The broadcasting rate is low if a high-frequency band is chosen, as these frequencies do not penetrate through the walls, buildings, etc. And, if the operating frequency is very small, then a large antenna is required, which is not feasible for practical reasons. For example, 0.9 GHz is also an ISM band and can easily broadcast the signal to a longer distance, but the antenna size is large as compared to the 2.4 GHz.

1.6 Recent Developments

In this section, a review of new emerging technologies used for rectenna design is presented. WPT research was first demonstrated by Brown [43]. Multiple designs of the rectenna components are observed like variation of antennas, for example single band to multiband or broadband using different techniques like slit/slot, DGS, reconfigurable, circular polarization, etc. [44–46]. Different topologies of rectifiers are used with variations in the circuit of impedance matching to satisfy the condition of MPT. Wireless electricity seems like an excellent idea, but while designing its transceiving system, the researchers face many challenges such as less conversion efficiency, components mismatching, etc. [47]. In this section, the recent techniques employed by the researchers to encounter such issues are presented.

In [48], a dual-band dual-polarized rectenna was presented, which harvests the RF energy at 2.4 and 5.5 GHz. Two microstrip feeding lines were used with a 180° phase shift as shown in Fig. 1.7. The antenna was fabricated on the Teflon of thickness of 0.8 mm. The efficiency of the antenna was more than 9% at both bands.

In [49], a broadband, flat dipole antenna covering the ISM band was reported. A rectangular electromagnetically coupled resonator was used, which reduces the overall size of the antenna. Here, two configurations of the rectenna were presented. In the first type, the rectifier was connected orthogonally with the antenna, while in the second type, the rectifier was connected parallelly with the antenna as shown in Fig. 1.8.

To make the antenna more compact and efficient, D. B. Lin had presented an antenna pair for WPT. The transmitting and receiving antennas were integrated into a single unit to achieve maximum conversion efficiency. The transmitting antenna was PIFA, while the receiving was microstrip antenna. The gap between the two antennas is 7 mm, and the air gap is 2.8 mm. This antenna system is applicable for far-field and as well as for near-field wireless transmission.

In [50], a reconfigurable rectifier was presented with a load resistance R_L of 220 Ω. A wide input power range was achieved at 5.8 GHz using a shorted stub

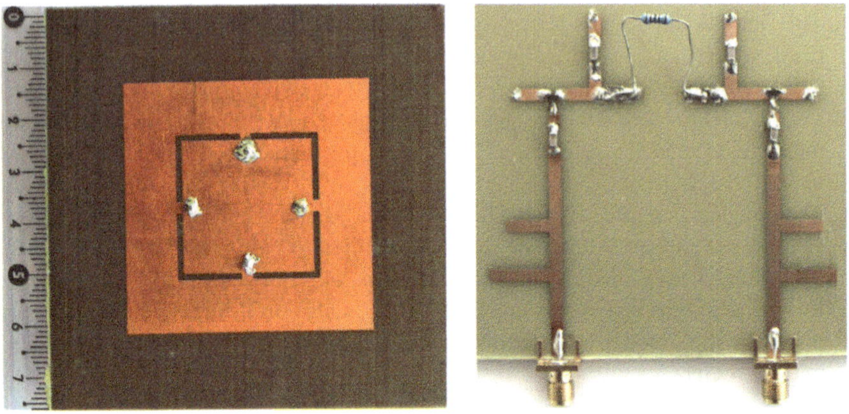

Fig. 1.7 Dual-band dual-polarized rectenna [48]

(a) (b)

Fig. 1.8 ISM band rectenna. **a** Orthogonal connection and **b** parallel connection [49]

and the PIN diode. A good impedance matching circuit was designed with an L-shaped reconfigurable stub to maximize the conversion efficiency, shown in Fig. 1.9. The PIN diode was used to connect/disconnect the L-shaped stub to the circuit by switching the diode to ON/OFF.

The above discussion on the rectenna is limited to the microwave range, but several researchers are working in the millimeter and terahertz range also [51]. At the microwave range, the Schottky diode is the most popular choice for rectification due to its multiple advantages discussed in the subsequent chapters. But, at millimeter range, the Schottky diode shows a very narrow band and extremely low conversion

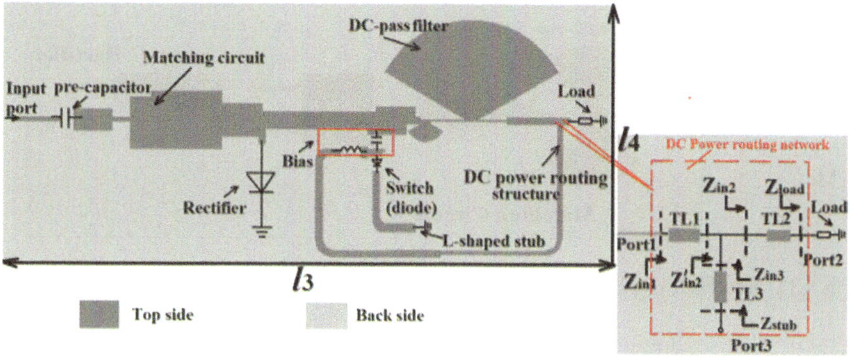

Fig. 1.9 Reconfigurable rectifier [50]

efficiency. Different techniques are being implemented to replace the Schottky diode to increase conversion efficiency.

In [52], a rectifier using CMOS technology was reported. A common gate and four input comparators were used to derive an NMOS power switch to match a zero-threshold diode. The comparator used in designing the circuit is a 2 ns voltage comparator. This circuit was fabricated by using a 0.35 μm CMOS process. This circuit achieved the maximum DC voltage of 3.22 V and was implemented for biomedical applications.

In [53], an RF energy harvester working in the UHF band was reported. It showed high power conversion efficiency. The circuit was fabricated with the 65 nm CMOS technology as shown in Fig. 1.10. The system showed applications in biomedical

Fig. 1.10 CMOS fabricated rectifier [53]

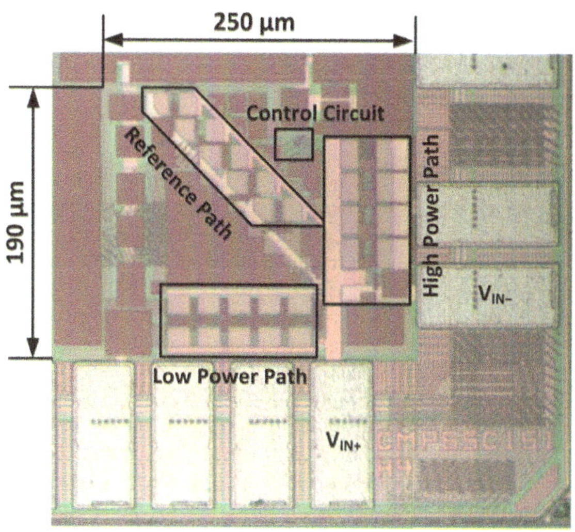

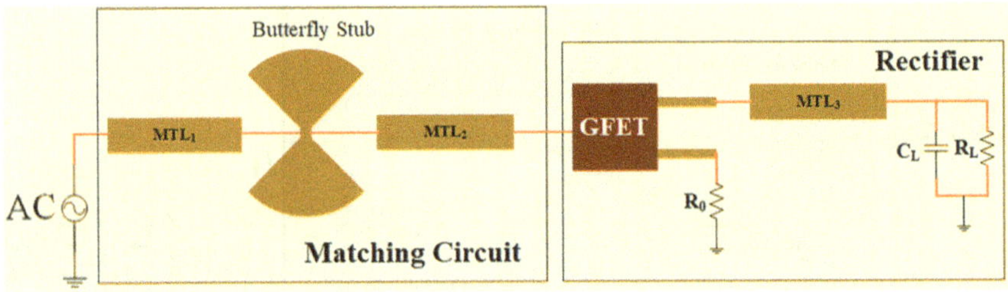

Fig. 1.11 Schematic of the graphene FET rectifier [55]

implantable devices, Internet of Things (IoT), wearable gadgets, and radio frequency identification (RFID).

The CMOS technology faces leakage current problem due to its high PCE. Large transistors are required to achieve high sensitivity, which, in turn, increases the reverse leakage current. Nowadays, the researchers are showing more interest in graphene-based technology due to the reverse leakage current problem of the CMOS. In [54], a GFET rectifier integrated monopole antenna was reported, which covers the range of 22.5–27.5 GHz. A compact and simple quarter-wave transformer was used to match the impedance of the antenna and the rectifier. The conversion efficiency of 80% was achieved at 5 dBm.

In [55], a graphene FET-based rectifier was reported for RF energy harvesting. The millimeter-wave rectifier showed working in the range of 29–46 GHz. A schematic of the graphene FET rectifier is shown in Fig. 1.11. The graphene shows multiple advantages, as compared to other semiconductor devices, in terms of stability, mobility, and saturation velocity. Therefore, it is considered as the best candidate for high-frequency rectification.

References

1. Khang S-T, Yu JW, Lee W-S (2015) Compact folded dipole rectenna with RF-based energy harvesting for IoT smart sensors. Electron Lett 51:926–928
2. Chuma EL, Rodrı́guez LDLT, Iano Y, Roger LLB, Sanchez-Soriano MA (2018) Compact rectenna based on a fractal geometry with a high conversion energy efficiency per area. IET Microw Antennas Propag 12:173–178
3. Sogorb T, Llario JV, Pelegrı́ J, Lajara R, Alberola J (2008) Studying the feasibility of energy harvesting from broadcast RF station for WSN. In: IEEE international instruments and measurement technology conference, Canada
4. Chen RH, Lee YC, Sun JS (2008) Design and experiment of a rectifying antenna for 900 MHz wireless power transmission. In: IEEE Asian-Pacific microwave conference, IEEE, Macau
5. Hong SSB, Ibrahim RB, Khir MHM, Zakariya MAB, Daud H (2013) Wi-Fi energy harvester for low power RFID application. Prog Electromagnet Res C 40:69–81
6. Shinohara N (2011) Be am efficiency of wireless power transmission via radio waves from short range to long range. J Korean Inst Electromagn Eng Sci 10(4):224–230

7. Ghosh S, Chakrabarty A (2016) Green energy harvesting from ambient RF radiation. In: 2016 international conference on microelectronics, computing and communications (MicroCom), Durgapur, pp 1–4
8. Lumpkins W (2014) Nikola Tesla's dream realized: wireless power energy harvesting. IEEE Consum Electron Mag 3(1):39–42
9. Ladou J, Lovegrove S (2008) Export of electronics equipment waste. Int J Occup Environ Health 14(1):1–10
10. Chen HW (2007) Exposure and health risk of gallium, indium, and arsenic from semiconductor manufacturing industry workers. Bull Environ Contam Toxicol 78(1):5–9
11. Grover P, Sahai A (2010) Shannon meets Tesla: wireless information and power transfer. In: Proceedings of IEEE ISIT, pp 2363–2367
12. Kang SH, Jung CW (2017) Textile resonators with thin copper wire for wearable MR-WPT system. IEEE Microwave Wirel Compon Lett 27(1):91–93
13. Chen Y, Xiao W, Guan Z, Zhang B, Qiu D, Wu M (2019) Nonlinear modeling and harmonic analysis of magnetic resonant WPT system based on equivalent small parameter method. IEEE Trans Industr Electron 66(8):6604–6612
14. Zhang J, Huang Y, Cao P (2015) A microwave wireless energy harvesting system with a wideband antenna array. Trans Inst Meas Control 37(8):961–969
15. Kazem Pour SM, Forghani M, Babakhani A (2020) Micrometer-sized sensors with free-space optical energy harvesting in CMOS. In: 2020 IEEE 20th topical meeting on silicon monolithic integrated circuits in RF systems (SiRF), San Antonio, TX, USA, pp 9–12
16. Matsumoto H (2002) Research on solar power station and microwave power transmission in Japan: review and perspectives. IEEE Microwave Mag 36–45
17. Shinohara N (2011) Power without wires. IEEE Microw Mag 12(7):S64–S73
18. Talla V, Pellerano S, Xu H, Ravi A, Palaskas Y (2015) Wi-Fi RF energy harvesting for battery-free wearable radio platforms. In: IEEE international conference, pp 47–54
19. Liu L, Zhang R, Chua KC (2013) Wireless information and power transfer: a dynamic power splitting approach. IEEE Trans Commun 61(9):3990–4001
20. Liu J, Xiong K, Fan P, Zhong Z (2017) RF energy harvesting wireless powered sensor networks for smart cities. IEEE Access 5:9348–9358
21. Song C, Lu P, Shen S (2020) Highly efficient omnidirectional integrated multi-band wireless energy harvesters for compact sensor nodes of internet-of-things. IEEE Trans Ind Electron https://doi.org/10.1109/TIE.2020.3009586
22. Wagih M, Hilton GS, Weddell AS, Beeby S (2020) Broadband millimeter-wave textile-based flexible rectenna for wearable energy harvesting. IEEE Trans Microw Theory Tech 68(11):4960–4972
23. Ladan S, Ghassemi N, Ghiotto A, Wu K (2013) Highly efficient compact rectenna for wireless energy harvesting application. IEEE Microwave Mag 14(1):117–122
24. Robinson BH (2009) E-waste: an assessment of global production and environmental impacts. Sci Total Environ 408(2):183–191
25. Hussain M, Mumtaz S (2014) E-waste: impacts, issues and management strategies. Rev Environ Health 29(1–2):53–58
26. Peter T, Rahman TA, Cheung S, Nilavalan R, Abutarboush HF, Vilches A (2014) A novel transparent UWB antenna for photovoltaic solar panel integration and RF energy harvesting. IEEE Trans Antennas Propag 62(4):1844–1853
27. Bryans AG, Fox B, Crossley PA, O'Malley M (2005) Impact of tidal generation on power system operation in Ireland. IEEE Trans Power Syst 20(4):2034–2040
28. Wang Y, Zhang Q, Zhao L, Tang Y, Shkel A, Kim ES (2016) Vibration energy harvester with low resonant frequency based on flexible coil and liquid spring. Appl Phys Lett 109:203901
29. Rozgic D, Markovic D (2017) A miniaturized 0.78-mW/cm^2 autonomous thermoelectric energy-harvesting platform for biomedical sensors. IEEE Trans Biomed Circuits Syst 11(4):773–783
30. Hemour S, Wu K (2014) Radio-frequency rectifier for electromagnetic energy harvesting: development path and future outlook. Proc IEEE 102(11):1667–1691

31. Harrington RF (1958) On the gain and beamwidth of directional antennas. IRE Trans Antennas Propag 6(3):219–225
32. Talla V, Kellogg B, Ransford B, Naderiparizi S, Gollakota S, Smith JR (2015) Powering the next billion devices with Wi-Fi. In: ACM conference on emerging networking experiments and technologies
33. Valenta CR, Durgin GD (2014) Harvesting wireless power: survey of energy-harvester conversion efficiency in far-field wireless power transfer systems. IEEE Microw Mag 15(4):108–120
34. Environmental assessment for the satellite power system-concept development and evaluation program-microwave health and ecological effects (1980). https://doi.org/10.2172/6736376
35. David L (1980) Study of federal microwave standards. https://doi.org/10.2172/5021571
36. Jayaswal G, Belkadi A, Meredov A, Pelz B, Moddel G, Shamim A (2018) Optical rectification through an Al_2O_3 based MIM passive rectenna at 28.3 THz. Mater Today Energy 7:1–9
37. Jayaswal G, Belkadi A, Meredov A, Pelz B, Moddel G, Shamim A (2018) A zero-bias, completely passive 28 THz rectenna for energy harvesting from infrared (waste heat). In: 2018 IEEE/MTT-S international microwave symposium—IMS, Philadelphia, PA, pp 355–358
38. Jiang Z, Zhang P, Huang L, Zhang J, He X, Rihan M (2020) Lens antenna arrays aided co-existing radar and communication systems with energy harvesting. IEEE Access 8:56160–56169
39. Estrada J, Ramos I, Narayan A, Keith A, Popovic Z (2016) RF energy harvester in the proximity of an aircraft radar altimeter. In: 2016 IEEE wireless power transfer conference (WPTC), Aveiro, pp 1–4
40. Muhammad Zin N, Mohamed Jenu MZ, Ahmad Po'ad F (2011) Measurements and reduction of microwave oven electromagnetic leakage. In: 2011 IEEE international RF & microwave conference, Seremban, Negeri Sembilan, pp 1–4
41. Soltysiak M, Celuch M, Erle U (2011) Measured and simulated frequency spectra of the household microwave oven. In: 2011 IEEE MTT-S international microwave symposium, Baltimore, MD, pp 1–4
42. Risman PO, Celuch-Marcysiak M (2000) Electromagnetic modelling for microwave heating applications. In: 13th international conference on microwaves radar and wireless communications MIKON 2000, vol 3, pp 167–182, 22–24 May 2000
43. Brown WC (1983) Design study for a ground microwave power transmission system for use with a high-altitude powered platform. NASA Final Contractor Rep 168344, Raytheon Rpt. PT-6052, NASA Contr. NAS6–3200
44. Kwak SI, Kwon JH, Sim D, Chang K, Yoon YJ (2010) Design of the printed slot antenna using wiggly line with harmonic suppression. IEEE Antennas Wirel Propag Lett 9:741–743
45. Lou X, Yang G (2018) A dual linearly polarized rectenna using defected ground structure for wireless power transmission. IEEE Microwave Wirel Compon Lett 28(9):828–830
46. Sharma A et al (2019) Wideband high-gain circularly-polarized low RCS dipole antenna with a frequency selective surface. IEEE Access 7:156592–156602
47. Boaventura AS, Carvalho NB (2011) Maximizing dc power in energy harvesting circuits using multisine excitation. In: IEEE MTT-S international microwave symposium digest 2011
48. Mattsson M, Kolitsidas CI, Jonsson BLG (2018) Dual-band dual-polarized full-wave rectenna based on differential field sampling. IEEE Antennas Wirel Propag Lett 17(6):956–959
49. Okba A, Takacs A, Aubert H (2019) Compact rectennas for ultra-low-power wireless transmission applications. IEEE Trans Microw Theory Tech 67(5):1697–1707
50. Lu P, Song C, Cheng F, Zhang B, Huang K (2020) A self-biased adaptive reconfigurable rectenna for microwave power transmission. IEEE Trans Power Electron 35(8):7749–7754
51. Khan AA, Jayaswal G, Gahaffar FA, Shamim A (2017) Metal-insulator-metal diodes with sub-nanometre surface roughness for energy-harvesting applications. Microelectron Eng 181:34–42
52. Lam Y, Ki W, Tsui C (2006) Integrated low-loss CMOS active rectifier for wirelessly powered devices. IEEE Trans Circuits Syst II Express Briefs 53(12):1378–1382
53. Lu Y et al (2017) A wide input range dual-path CMOS rectifier for RF energy harvesting. IEEE Trans Circuits Syst II Express Briefs 64(2):166–170
54. Singh N, Kumar S, Kanaujia BK et al (2020) A compact broadband GFET based rectenna for RF energy harvesting applications. Microsyst Technol 26:1881–1888
55. Singh N (2020) A compact and efficient graphene FET based RF energy harvester for green communication. Int J Electron Commun (AEÜ) 115:153059–153066

Chapter 2
Background and Origin of the Rectenna

2.1 History of the Rectenna

In recent times, renewable energy sources are becoming a need to energize low-power electronic devices. Rectenna offers a continuous power supply by harvesting the EM waves available in the environment and converting them into DC power [1, 2]. A rectenna is location independent and pollution-free and does not depend on the climatic conditions [3]. The primary design goal for the rectenna is to achieve maximum conversion efficiency so that sufficient power can be provided to the electronic gadgets. Many other design factors are also considered such as cost, availability, and size of the device. This chapter gives an overview of the rectenna and its advancements. Also, different methods have been discussed to achieve high conversion efficiency.

J.C. Maxwell was the first scientist to give the theoretical existence of EM waves [4]. H. Hertz further carried Maxwell's work and proved the existence of the EM wave in 1887 [5]. He designed a loop antenna that can transmit and detect the radio waves.

After Maxwell and Hertz, N. Tesla demonstrated that wireless systems could provide electricity, as shown in Fig. 2.1. Tesla first gave the concept of wireless transmission of power in 1897 [6, 7] and performed his experiment at Colorado Lab in 1899. He designed a coil of 16 m diameter that resonates at a frequency of 150 kHz [8]. This experiment was carried at the top of the Wardenclyffe Tower, mentioned in Fig. 2.2. In the early 1930s, H.V Nobel designed the transmitting and receiving dipoles, which were kept 25 ft apart at 100 MHz as shown in Fig. 2.3. However, this experiment was failed. The main reason for failure was the direction of radio waves. The radio waves were not concentrated on a particular path. It was only after World War II, several researchers showed interest in wireless power transfer systems. An improved microwave power transfer (MPT) technology was demonstrated in the Raytheon Lab., NASA, in 1958, shown in Fig. 2.4. W.C. Brown proposed the concept of rectenna in 1963 [9]. He obtained a conversion efficiency of 50% by his first-ever

B. K. Kanaujia et al., *Rectenna: Wireless Energy Harvesting System*,
Advances in Sustainability Science and Technology,
https://doi.org/10.1007/978-981-16-2536-7_2

Fig. 2.1 N. Tesla demonstrated the wireless power transmission [7]

designed 'rectenna.' This experiment was performed at 2.45 GHz using the point-contact diode. They further demonstrated directive devices with reflectors and horns [10].

In 1964, Brown solved the problem of low conversion efficiency by using an impedance matching network between the antenna and the rectifying diode [11]. An array of rods was used as a matching network. Figure 2.5 shows the helicopter rectenna constructed by using 448 IN82G diodes, and it was able to achieve a power of 270 W, which can operate the rotor of the helicopter. They flew the helicopter for more than 10 h at a height of 60 ft. This marks the beginning of the 'rectenna' [12].

In 1968, HP demonstrated that the Schottky diode has a better power handling capability and low threshold voltage, as compared to the point-contact diodes [13, 14]. The Schottky diode improved the performance of the rectenna array, and 20 W DC power was reported by using a full-wave bridge rectifier. However, the low conversion efficiency was still a problem, and in 1975, an efficiency of 54% was achieved at the Raytheon Lab using magnetrons [15].

2.1.1 Development in the Field of Solar Power Satellite

In 1968, P. Glaser introduced the concept of solar power satellite (SPS), where solar energy is collected and converted into electrical power. Microwave transmitters at the satellite convert this electricity into the radio frequency waves and transmit it to the earth. The radio waves are received by a rectenna located at the earth [16]. As compared to the ground-based power transmission, SPS offers high conversion efficiency due to the easy availability of solar energy [17].

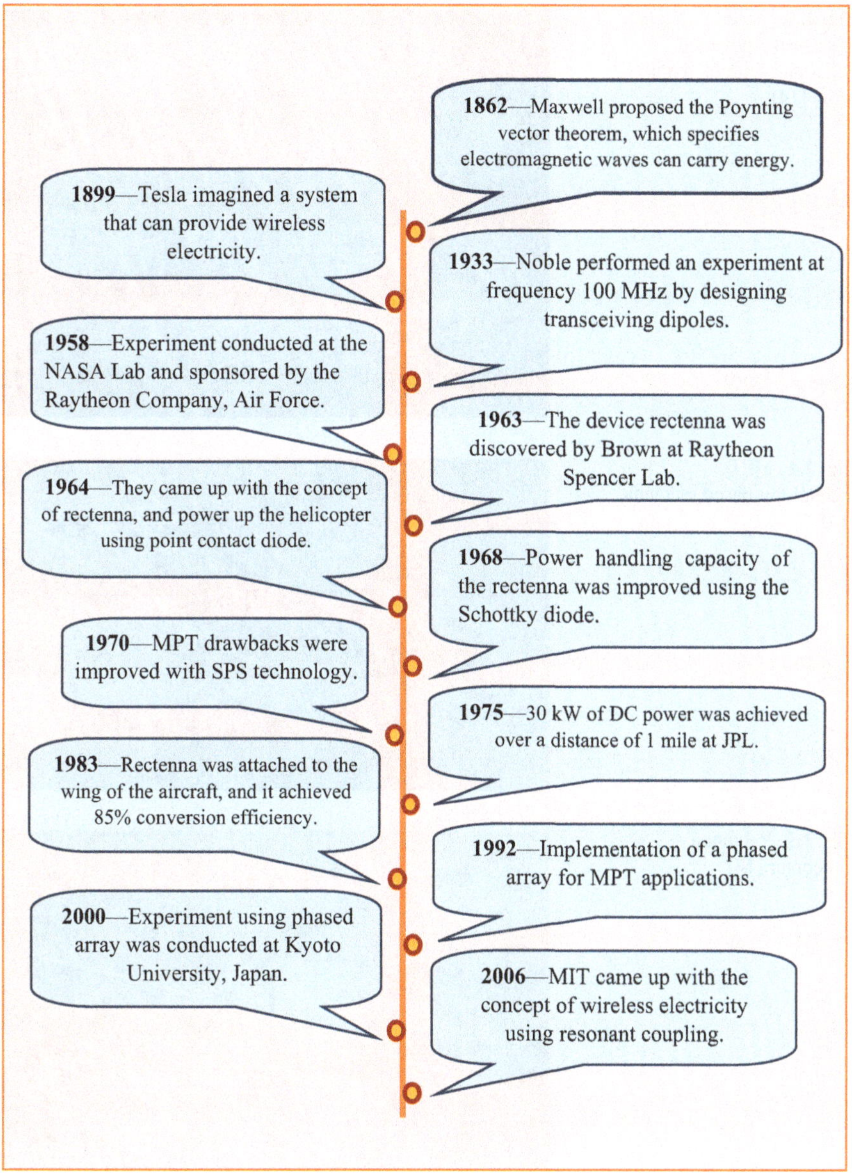

Fig. 2.2 Various experiments performed on the rectenna in past decades

Fig. 2.3 Microwave-powered helicopter with rectenna array [14]

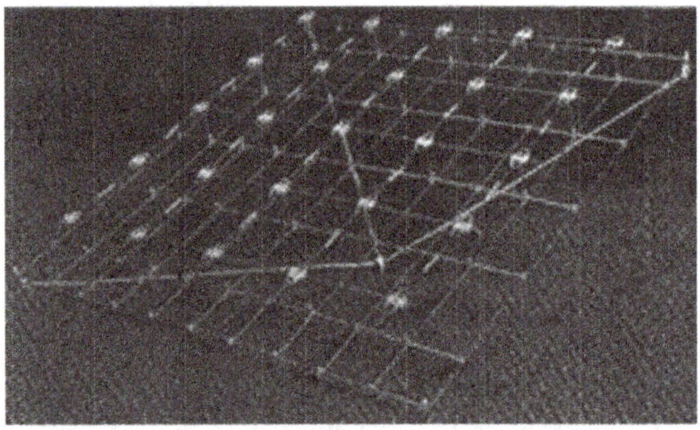

Fig. 2.4 MPT technology-based rectenna [14]

Fig. 2.5 Rectenna helicopter [14]

SPS WPT technology has many advantages over solar cells. Solar cells depend on the climatic conditions; therefore, their efficiency is low. In SPS, the converted power is 10 times more than the solar cells as microwaves do not depend on the weather. There is no need for expensive storage devices as sunlight is easily available in the space and the dissipated heat is also radiated back into space. In 1983, an SPS-based microwave ionosphere nonlinear interaction experiment (MINIX) was conducted in Japan [18, 19] as shown in Fig. 2.6. In this experiment, an antenna array comprised

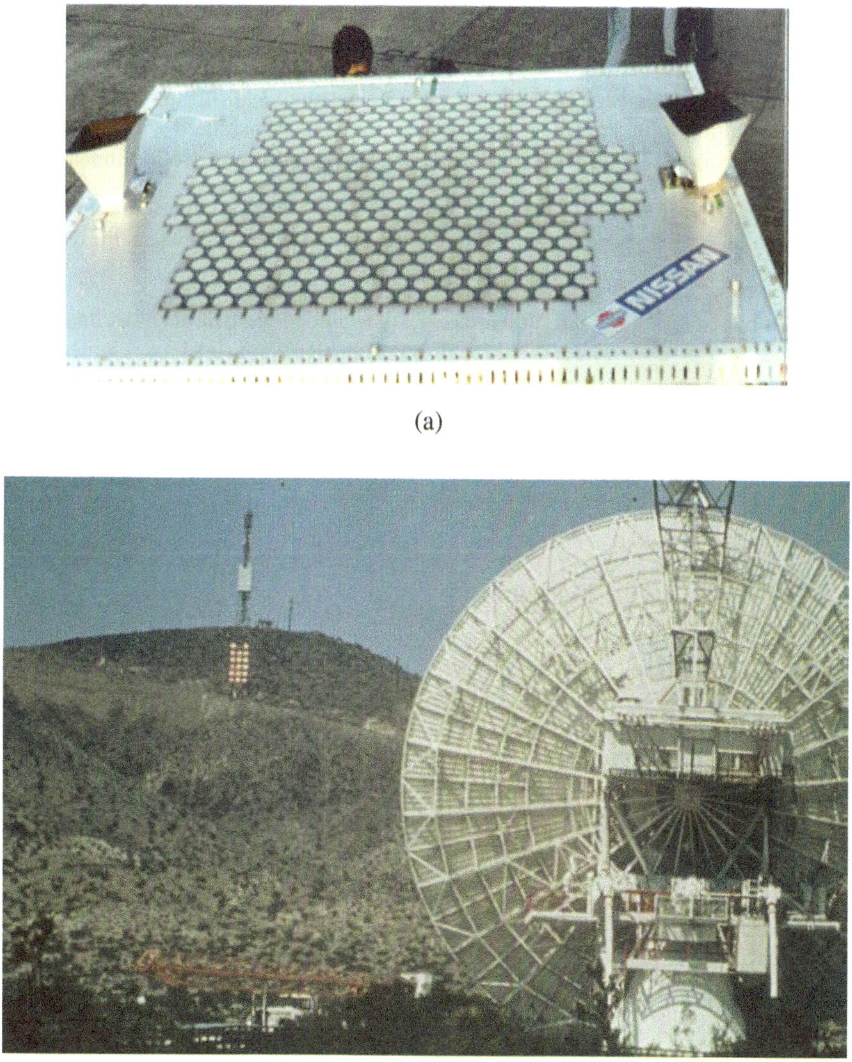

(a)

(b)

Fig. 2.6 **a** Microwave-powered phased array airplane, **b** ground-to-ground MPT [18]

of 288 elements was used to run a fuel-free helicopter. In 1995, NASA carried out various experiments on SPS technology. A power of 20 W was achieved through a phased array from a distance of 150 km [20, 21]. However, this technology was not successful due to various issues, such as large sizes of the equipment, maintenance of solar panels in space is difficult, and interference may be possible with other communication satellites [22].

2.2 Rectenna Technology

Clean energy is a need for present and future electronic systems. WPT technology can be a solution to this problem, where power will be transmitted without wires. It also has less impact on human health [23, 24]. WPT follows the following steps:

- The RF power is sensed by the sensing device that operates at the same frequency.
- The collected RF power is converted into DC to provide power supply to the electronic devices or charging batteries.

Providing electricity to the rural areas is still a challenge, in terms of technical and economic factors [25]. WPT can solve this difficulty. Rectenna is a basic element used to harvest RF energy available in the environment. It consists of a receiving antenna, rectifier, and impedance matching circuit as shown in Fig. 2.7. The rectenna designs reported in the open literature showed single, multiple, and broadband operation, shown in Fig. 2.8 [26–28].

2.2.1 *Single-Band Rectenna*

A single-band rectenna resonates at one frequency. Various configurations of single-band rectenna have been reported in the literature. In [29], a rectenna consisted of a circularly polarized microstrip antenna was presented, where an unbalanced circular slot was loaded on the sensing patch to suppress higher-order harmonics. A circular

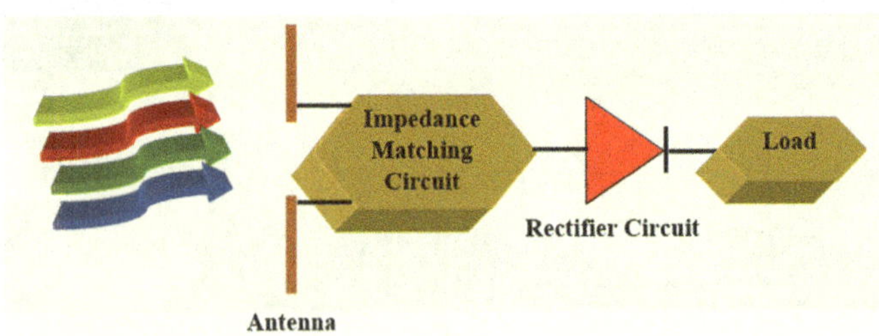

Fig. 2.7 Block diagram of the rectenna

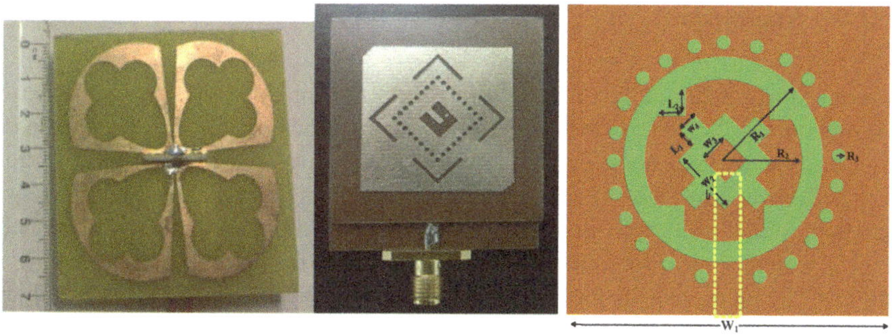

Fig. 2.8 Single-band, multiband, and wideband antenna prototypes [26–28]

rectifying antenna was reported with triangular-shaped slots [30]. A circularly polarized antenna with truncated corners, and a U-shaped slot embedded in the patch, was proposed [31].

2.2.2 Broadband Rectenna

The broadband rectenna covers a wide range of frequencies; therefore, its efficiency is high. The impedance bandwidth of the sensing antenna must be large to design a broadband rectenna [32]. A rectenna with slotted ground plane showed wide bandwidth [33]. In [34], a compact size Vivaldi rectenna was proposed for broadband characteristics.

2.2.3 Multiband Rectenna

Multiband rectenna functions at multiple frequency bands such as GSM, WLAN, and Wi-MAX. Since it is difficult to match impedance for a wide range of frequencies, the multiband rectenna shows high conversion efficiency.

The impedance matching circuit is designed to match all the frequencies to obtain high conversion efficiency. Limited works are reported on the multiband rectenna with high conversion efficiency [35–40]. A triple-band planar inverted-F antenna (PIFA) integrated with a meandered line was reported [35]. A rectenna consisted of pentagon DRA, and a rectangular slot on the ground plane was reported for dual-band performance [36]. Various techniques were proposed to achieve multiband characteristics like the air gap [37], fractal shape [38], differential feeding [39], aperture coupled [40], etc.

2.2.4 Rectenna Array

Various single-band, multiband, and broadband rectenna designs were reported in the literature. All such reported rectenna designs consisted of a few sensing antennas, which are sometimes not efficient to collect adequate power. It is useful for long-distance power transfer as more power is collected by numerous elements connected together. A high gain was achieved with a log-periodic dipole antenna array reported in [41], which increases the conversion efficiency of the rectenna. The received power increases by increasing the feeding lines in a rectenna array [42]. In [43], a multiport antenna was implemented with the tilted beam grid array to reduce the size of the rectenna. The converted DC output power from different elements of the array can be combined in hybrid, series, or parallel [44, 45].

2.2.5 Optical Rectenna

An optical rectenna converts light energy into DC power. It can be designed using nanotubes that act as an optical sensor and capture energy from the sunlight or other light sources. In optical rectenna, the main challenge is to match impedance between the antenna and the rectifier [46]. Fabrication is also one of the problems at nanometers. A bow-tie nano-antenna was reported to convert light into DC power, where rectification was carried out using a tunnel diode [47]. Optical rectenna using quantum technology was presented in [48], which directly converted solar energy to DC power.

2.2.6 Rectenna Architecture

Rectenna architecture is mainly comprised of three components: the RF energy source, information gateway, and the network nodes. A generalized RF energy harvesting architecture is shown in Fig. 2.9. The information gateway, shown in the wireless network, consists of cellular base stations and relays. This system provides energy to Wi-Fi routers, mobile phones, TV towers, and laptops to radiate EM signals [49–52]. In some cases, the RF source and information gateway are the same, called infrastructure-less architecture as shown in Fig. 2.9b. In infrastructure-based architecture, shown in Fig. 2.9a, a base station or access point is required to operate a wireless network, while the infrastructure-less architecture does not require any access point or base station, and it is based on peer-to-peer connection. From the information gateway, the power is supplied to the RF sources that radiate microwave power in the environment. The network node senses the radiated microwave signals and converts them into DC power. The DC power can be used directly or stored in the storage devices for later use.

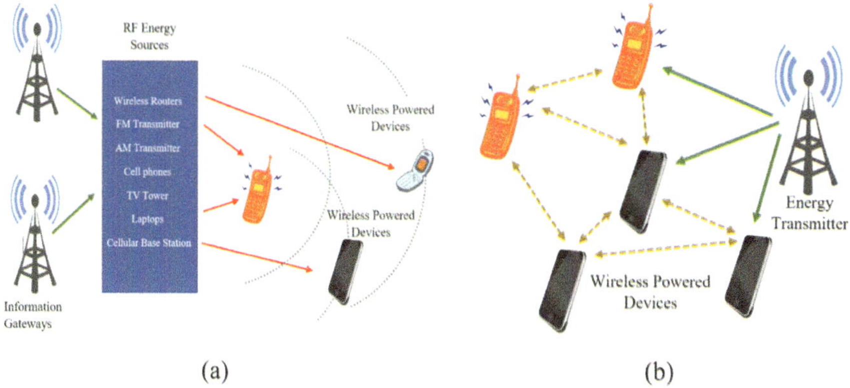

Fig. 2.9 Rectenna architectures. **a** Infrastructure based and **b** infrastructure-less

This system requires the following components to harvest RF energy.

- A source also called the main supply.
- RF transceivers that can receive signals from the base stations and transmit them to energy harvesting devices.
- Rectenna that is used to convert EM waves into the DC power.
- A power management system that controls the DC power. The power can be controlled using the following methods:

Harvest use: In this module, the DC power can be directly used to supply energy or recharge the batteries.

Harvest store use: In this module, an energy storage device is used to store DC power for later use. This approach is also used when node power exceeds a specified limit.

2.3 Types of WPT

WPT is broadly classified as near-field WPT and far-field WPT. Both techniques work on the principle of electromagnetic power transmission [53, 54]. Figure 2.10 shows the classification of WPT techniques. The classification mainly depends on the separation distance between the receiver and transmitter. Near-field WPT is also called as non-radiative coupling-based power transmission, and far-field is also called as radiated power transmission. The receiving device captures the propagating EM waves and converts them into DC power.

In the non-radiative method, the power is transferred due to the coupling between the two coils, which may be inductive or capacitive. In the radiative power transmission, the beamforming technique is used by means of microwaves or laser beam. The

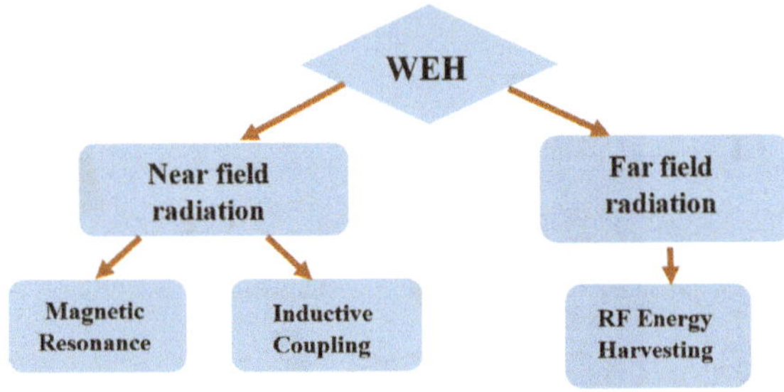

Fig. 2.10 Classification of different types of WPT techniques

attenuation of RF signal from transmitter to receiver is inversely proportional to the separation distance.

2.3.1 Near-Field WPT

The near-field WPT has a limited distance range, which is about one wavelength of the transmitting device. Therefore, it is also known as short-wave transmission. In this method, the time-varying electric field generates a time-varying magnetic field. The power transferred through the magnetic field is called inductive coupling, and the power transferred through the electric field is called capacitive coupling [55]. The transmitting power leaves the transmitter if the receiving or the absorbing material is present in the specified range. The power received at the receiver decreases exponentially with the distance.

2.3.1.1 Inductive Coupling

This configuration consists of two coils, named as primary and secondary, as shown in Fig. 2.11. The magnetic field generated by a time-varying electric field induces a current in the secondary coil. A conducting material is used to design the primary coil. The primary coil develops an oscillating electric field and generates a time-varying magnetic field. The secondary coil detects the magnetic field generated by the primary coil. The power is transferred efficiently if the separation distance between the coils is less than a single wavelength. The induced magnetic field is responsible for the electric current flow, which may be used directly or can be stored in the energy storage devices like a capacitor or battery for future use.

In addition, the power received by the secondary coil depends on the orientation of the coil as shown in Fig. 2.11. In the first case, the secondary coil is parallel to the primary coil; hence, maximum power is received, which shows the maximum conversion efficiency. In the second case, the secondary coil is perpendicular to the primary coil; hence, the minimum power is received, which shows minimum conversion efficiency. This technique is simple and easy to implement as a transfer of

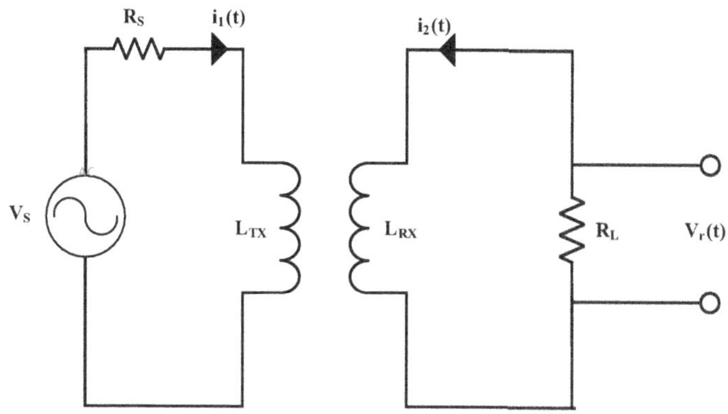

Fig. 2.11 Circuit diagram of inductive coupling WPT

energy is confined between the coils and no power is radiated to the near surroundings. Hence, it has various applications in the healthcare field, RFID tags, and the charging of mobile devices. This configuration can be used in different topologies like parallel–parallel, parallel–series, series–parallel, and series–series.

2.3.1.2 Magnetic Resonant Coupling

As compared to other techniques, this system transfers a high amount of power. Similar to the inductive coupling method, two coils (primary and secondary) are used in this method, and capacitors are also added to the circuit. The two transmitting and receiving circuit capacitors are added to the circuit that improves the performance of the system. The magnetic resonant coupling allows the charging of multiple devices concurrently. The mutual inductance (M) between the transmitting and receiving circuit can be expressed as

$$I_{\mathrm{R}}(R_{\mathrm{L}} + J\omega L_{\mathrm{R}} + 1/J\omega C_{\mathrm{R}} + Z_{\mathrm{R}}) = j\omega M I_{\mathrm{T}} \tag{2.1}$$

where I_{T} is the transmitting circuit, I_{R} is the receiving current, L_{R} is the inductance at receiving coil, R_{L} is the load resistance, C_{L} is the load capacitance, and Z_{L} is the complex impedance. The angular frequency of the system is calculated as

$$\omega = \frac{1}{\sqrt{LC}} = \frac{1}{\sqrt{L_{\mathrm{T}}C_{\mathrm{T}}}} = \frac{1}{\sqrt{L_{\mathrm{R}}C_{\mathrm{R}}}} \tag{2.2}$$

The voltage at the transmitter is expressed as

$$V_{\mathrm{T}} = I_{\mathrm{T}}(R_i + J\omega L_{\mathrm{T}} + 1/J\omega C_{\mathrm{T}} + Z_i) - j\omega M I_{\mathrm{T}} \tag{2.3}$$

This coupling scheme can have four different topologies: single-input single-output (SISO), multi-input single-output (MISO), single-input multi-output (SIMO), and multi-input multi-output (MIMO) as shown in Fig. 2.12. The received power for SISO configuration can be obtained as

$$P_R = P_T Q_T Q_R \eta_R \eta_T k^2(d) \tag{2.4}$$

where P_T is the transmitting power, Q_T and Q_R are quality factors of the transmitter and receiver, respectively, η_T is the transmitter efficiency, η_R is the receiver efficiency, k is the coupling coefficient, r_T, and r_R are the radii of transmitting and receiving coils, respectively, and d is the separation distance.

$$\eta_T = \frac{Z_S}{R_T + Z_S}, \eta_R = \frac{Z_L}{R_R + Z_L} \tag{2.5}$$

$$Q_T = \frac{\omega L_S}{R_T + Z_S}, \eta_R = \frac{\omega L_L}{R_R + Z_L} \tag{2.6}$$

The power received at the receiver with MISO configuration can be expressed as

$$P_R^n = P_T^n Q_T^n Q_R^n \eta_R^n \eta_T^n k_n^2(d) \tag{2.7}$$

where coupling coefficient can be obtained as

$$k_n^2(d) = \frac{r_n^3 r_R^3 \pi^2}{\left(d_n^2 + r_n^2\right)^3} \tag{2.8}$$

The total power from n transmitter coils is expressed as

$$P_R^n = Q_R \eta_R \sum_{n=1}^{N_T} P_T^n Q_T^n \eta_T^n k_n^2(d) \tag{2.9}$$

Similarly, the received power for SIMO is given by

$$P_R^m = P_T^m Q_T^m Q_R^m \eta_R^m \eta_T^m k_m^2(d) \tag{2.10}$$

The total power from n transmitter coils is expressed as

$$P_R^m = P_T Q_T \eta_T \sum_{m=1}^{N_R} Q_R^m \eta_R^m k_m^2(d) \tag{2.11}$$

And, the received power for MIMO is given by

$$P_R^{n,m} = P_T^n Q_T^n Q_R^m \eta_R^m \eta_T^n k_{n,m}^2(d) \tag{2.12}$$

The total power from n transmitter coils is expressed as

$$P_{\mathrm{R}} = \sum_{n=1}^{N_{\mathrm{T}}} \sum_{m=1}^{N_{\mathrm{R}}} P_{\mathrm{T}}^{n} Q_{\mathrm{T}}^{n} Q_{\mathrm{R}}^{m} \eta_{\mathrm{T}}^{n} \eta_{\mathrm{R}}^{m} k_{n,m}^{2}(d) \tag{2.13}$$

This configuration shows applications in the field of charging of portable devices like medical implants, laptops, electric vehicles, electric appliances, cell phones, etc.

- **Capacitive coupling**: Capacitive coupling is a conjugate of inductive coupling. In this method, the electrodes (such as metal plates) replace the coils, and they work like a capacitor. The space between the transmitter and receiver electrodes is called dielectric. The transmitting electrode generates a time-varying electric field, which induces a varying potential on the metal plate of the receiver. As the frequency and capacitance between the electrodes increases, the power transfer efficiency increases; however, it decreases with the increasing distance. This configuration has two methods:

Longitudinal: Only one electrode is used for power transfer, and it is terminated with a passive electrode or ground plane. The electric field oscillations are confined between the active and passive metal plates.

Transverse: This circuit comprised of two metal plates as the transmitter and two metal plates as the receiver. The oscillating electric field generated from the transmitter induces an opposite phase (180°) alternating potential on the receiver plate. Henceforth, the rectifier converts the alternating potential into DC voltage.

2.3.2 *Far-Field WPT*

The propagating electromagnetic wave consists of magnetic and electric fields that are perpendicular to each other. The electromagnetic waves can be in the form of microwaves, millimeter wave, and light waves [56, 57]. This method is applied for long-range applications when the distance is in kilometers. In this region, the square of distance is inversely proportional to the radiated power.

2.3.2.1 RF Energy Harvesting

The RF signal frequency varies from 300 MHz to 300 GHz [58]. The antenna connected at the receiver to receive EM signals is known as sensing antenna. This antenna can be directional or omnidirectional with high gain. The rectifier connected next to the antenna rectifies the EM signal to the DC. A capacitor in the rectifying circuit is used to remove ripples from the rectified signal and obtain a pure DC power. The ratio of the converted DC power and the received AC power is termed as conversion efficiency.

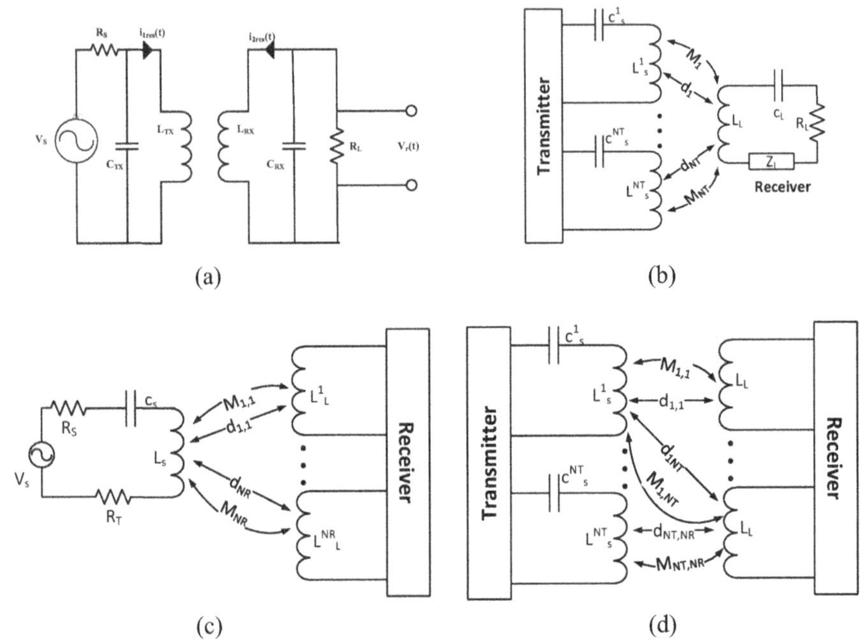

Fig. 2.12 **a** Magnetic resonance coupling, **b** multi-input single-output (MISO) coupling system, **c** single-input multi-output (SIMO) coupling system, **d** multi-input multi-output (MIMO) coupling system

$$\eta = \frac{P_{\mathrm{DC}}}{P_{\mathrm{AC}}} \tag{2.14}$$

The output DC power can be defined as

$$P_{\mathrm{DC}} = \frac{(V_{\mathrm{out}})^2}{R_{\mathrm{L}}} \tag{2.15}$$

The conversion efficiency is expressed as

$$\eta = \frac{(V_{\mathrm{out}})^2}{R_{\mathrm{L}} X P_{\mathrm{AC}}} \tag{2.16}$$

The RF energy harvesting mainly uses two methods for power transmission.

- **In-band RF energy transfer**: In this system, both transmitter and receiver resonate at the same frequency. Such systems are used for machine-to-machine communication, biomedical applications, and industrial applications.
- **Out-band RF energy transfer**: In this configuration, the operating frequency of the transmitter and the receiver are different. Such systems are used for satellite communication, where the uplink frequency range is different from the downlink

frequency range to avoid interference. These systems can be used for long-distance transmission.

2.4 Applications

A few applications of WPT are as follows:

2.4.1 *Charging of Vehicles*

Using MPT technology, electric vehicles can be charged automatically even while moving. This will save time and physical infrastructure, and it would also be beneficial during emergency situations. This can be implemented by constructing RF systems/sources close to the highways.

Figure 2.13 illustrates the architecture of the wireless charging of e-vehicles. A transmitter transmits the RF signal that could be captured by the receiving antenna fixed inside the vehicle, which converts RF into DC power.

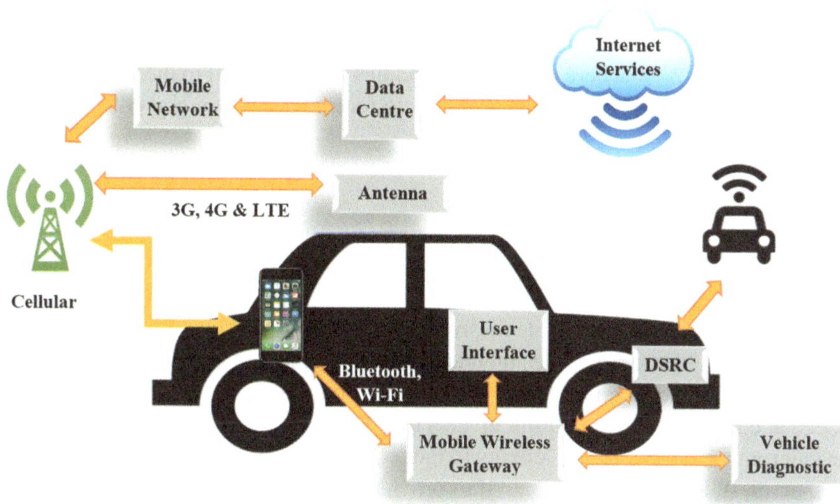

Fig. 2.13 Wireless charging of e-vehicles

Fig. 2.14 Wireless-powered home appliances using wi-tricity

2.4.2 *Self-sustainable Home Appliances*

The wirelessly powered appliances showed various advantages as compared to wired transmission. This may beneficial to the areas where wired connections are not feasible, like hilly and rural areas. Representation of a wireless-powered home is displayed in Fig. 2.14, where transmitter modules are connected to the walls, and RF chips are connected to the home appliances to receive RF power and convert it into DC power.

2.4.3 *Microwave-Powered Trains*

In the future, it could be possible to drive trains through microwaves radiations. As shown in Fig. 2.15, the receiving antenna connected at the rooftop of the train receives the RF signal and converts it into DC. In another method, a charging station can be placed at each station with a transceiver. This charging station receives energy from the environment and transmits energy to the receiver fitted on the rooftop of the train. Further, these RF radiations are used to provide electrical energy to run the train.

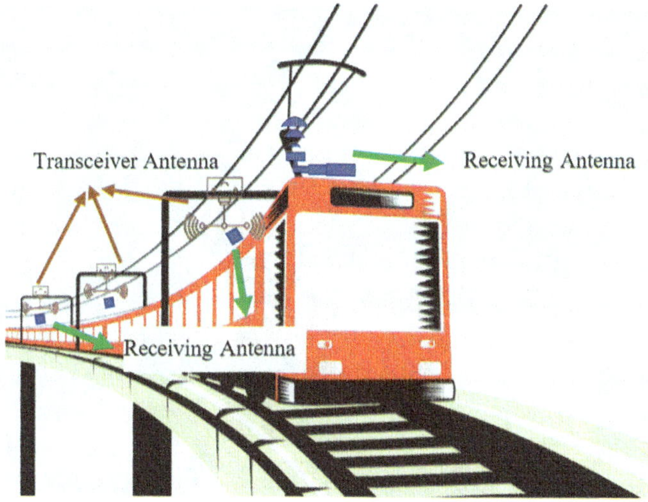

Fig. 2.15 Microwave-powered e-train

2.4.4 Wireless Drones

Drones are used for various day-to-day applications as they can easily reach the areas where humans cannot reach. The drones are powered through batteries, and therefore, they have less operation time, and battery needs to be changed frequently. This problem can be solved by using microwave power.

Figure 2.16 shows a picture of a firefighting vehicle that provides wireless power supply to the firefighting drones. The fire extinguishers use drones that are equipped with water pipes, and they are controlled by a remote control system installed in fire extinguisher vehicles. The transmitting source connected to the fire extinguisher vehicle transmits the EM waves, which could be sensed by the sensing antenna located on the drone. The sensing antenna receives the RF signal and converts it into DC power. The drones can also be used for military surveillance as wireless drones can easily send pictures/videos.

2.4.5 Smart Medical Health care

In the future, medical healthcare centers may also work on wireless electricity. A transmitter attached to the power station sends an EM signal that is received by a rectenna module. It converts it into an electrical signal and transmits wireless electricity to the medical devices or equipment as shown in Fig. 2.17. Since continuous monitoring is needed to examine a patient, therefore wireless electricity might play an important role in the hospitals.

Fig. 2.16 Firefighting vehicle provides wireless power supply to firefighting drones

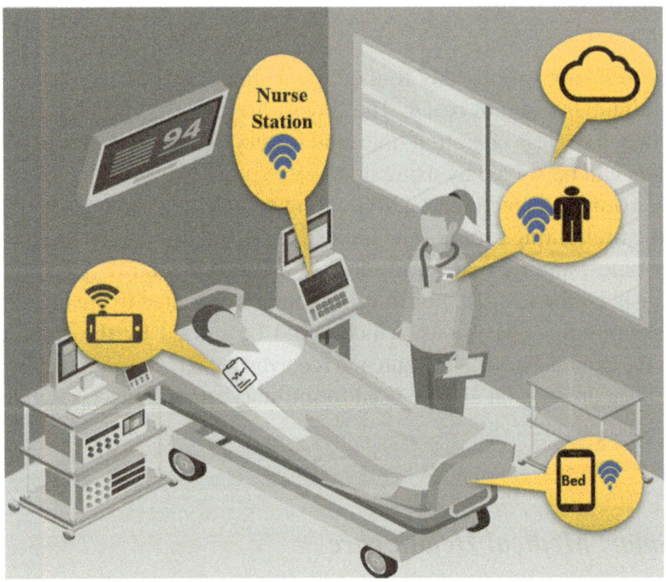

Fig. 2.17 Wireless-powered hospital equipments

Fig. 2.18 Wireless monitoring of the fields

The patient's health can be monitor by the medical staff or doctor using the implantable and wearable antennas powered through the wireless energy system. This technique is beneficial for the old-age patients who cannot visit the doctor regularly. The doctor can monitor their physical health remotely and prescribe the required medication.

2.4.6 *Smart Agriculture*

Smart farming also called intelligent farming controls or supervises the farm fields using different sensors. With the help of microwave sensors and the Internet of Things (IoT), farmers can track their fields using smart sensors as shown in Fig. 2.18. The water-level sensors can be used to determine the level of water required in the field. Soil sensors are used to detect the need for fertilizers, and the dielectric soil sensor monitors the moisture level of the soil.

2.4.7 *Wireless Power Grid*

Figure 2.19 shows a power transmission system where wireless power from the main station is supplied to the house. The power transmitted from the grid is converted into EM waves and transmit to the transmitting antenna. An antenna is used between the pathways to receive the signal and transmit it to the receiving antenna connected at the rooftop of the house. This system helps to provide electricity to rural areas.

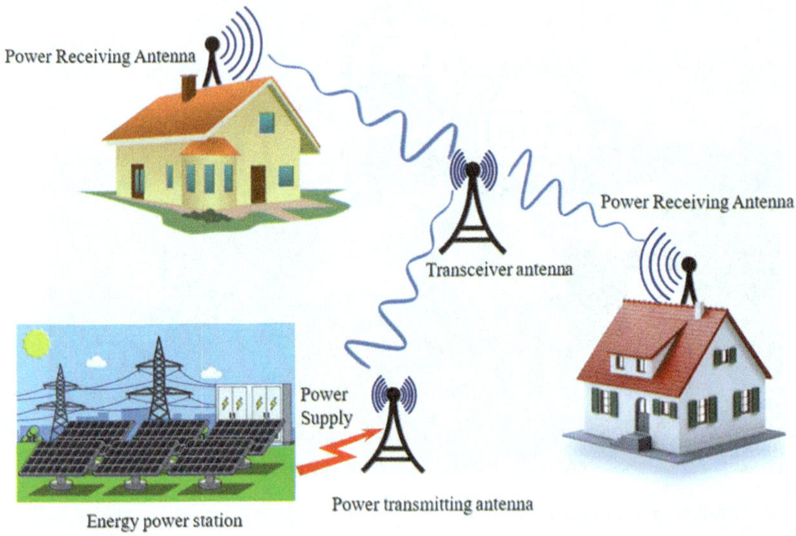

Fig. 2.19 Wireless power transmission from the base station to home

2.4.8 Smart City

In the future, clean and green power will be essential to run the digital world. In smart cities, as shown in Fig. 2.20, the data is collected by using IoT sensors and a number of operations are carried out to provide a better standard of living. The various actions are as follows:

- **Waste management**: continuous monitoring of the trash bins by the local authorities.
- **Smart education**: Parents can keep track of their children, and students can learn a lot of things with more clarity.
- **Smart energy**: Electricity maintenance is a big problem for the authorities. A smart system can manage power bills efficiently using the data provided by the sensor.
- **Health management**: A smart system can monitor the health of each resident living in the society/apartment from the health cards given to them.
- **Smart traffic management**: The vehicles can receive traffic information through a centralized server using IoT and WPT technology.

Fig. 2.20 World of wireless technology

2.4.9 *Self-driven e-Vehicles*

The forthcoming self-driving cars could move safely with the help of sensors and without human intervention. For proper functioning of the smart vehicles, there is a need to supply continuous power to the sensors, cameras, radar, and IoT devices. This continuous power can be provided by the rectenna through WPT as shown in Fig. 2.21.

2.4.10 *Microwave Power Sources in Disaster*

Different sensors and alarm systems can be installed at the bridges and forests to continuously monitor the water level and fire, respectively. The sensors and alarms need a continuous power supply for their operation. In the situation of a disaster, a medium is needed to maintain communication between the victims and their family members and to provide them food, clothes, first-aid kit, and other important items. The power can be provided with the help of airship connected to the transmit antenna, whose signals are captured by the group of drones. It consists of a transceiver that obtains a signal from the transmitting antenna and emits the EM waves into the environment. The receiving devices installed on the rooftop receive the power as shown in Fig. 2.22.

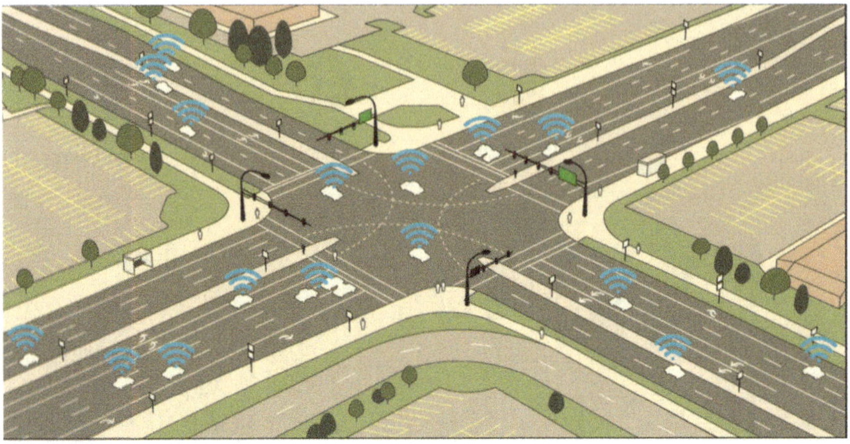

Fig. 2.21 Communication between e-vehicles

Fig. 2.22 Wireless technology in emergency situations

2.4.11 Medical Care of Animals

The animals cannot express their problems until the illness symptoms are shown on their body. It is a challenging task to discover their diseases and inform veterans to save their lives. As shown in Fig. 2.23, a device implanted in the body of animals can

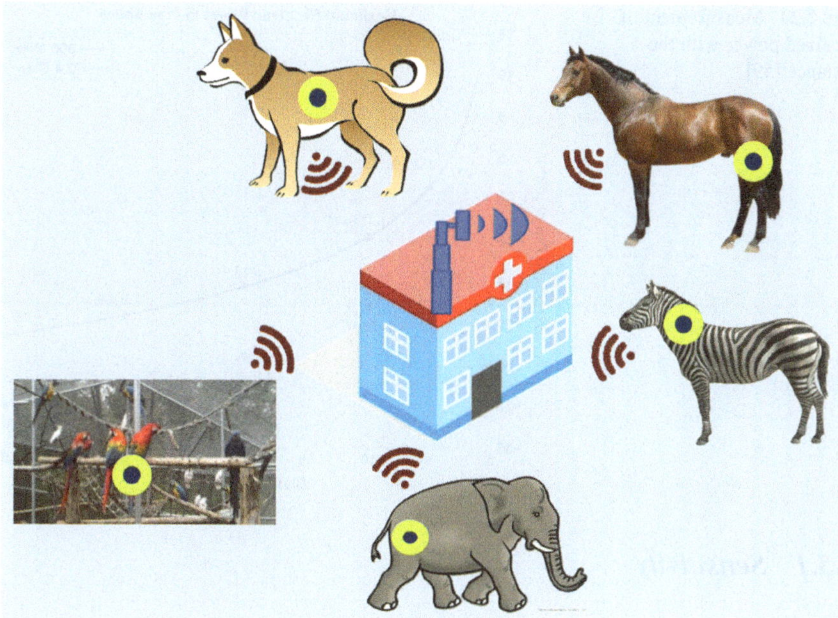

Fig. 2.23 Wireless health monitoring of the animals

solve this problem. With the help of this device, the medical staff can continuously monitor their health conditions and take necessary actions.

2.5 Power Available in the Ambient Environment

The power density harvested from natural or man-made sources is much low, which reduces the circuit performance and conversion efficiency. It is obvious that energy sources like solar, vibration, and thermal have the potential to generate more power, but, still, the RF sources are often used. RF energy harvesting system can provide enough energy to power the sensor nodes [59]. This system can withstand even in the harsh ambience where other energy resources cannot reach easily. In Fig. 2.24, the received power is shown at two different frequencies. At 40 m distance, maximum received power is 7 μW and 1 μW at 900 MHz and 2.4 GHz, respectively. This system also has some limitations, which can be resolved before its implementation [60].

To improve the system performance, some parameters need to be evaluated like sensitivity, operating range, resonating factor, separation distance, conversion efficiency, and output power, and their values can be varied according to the application.

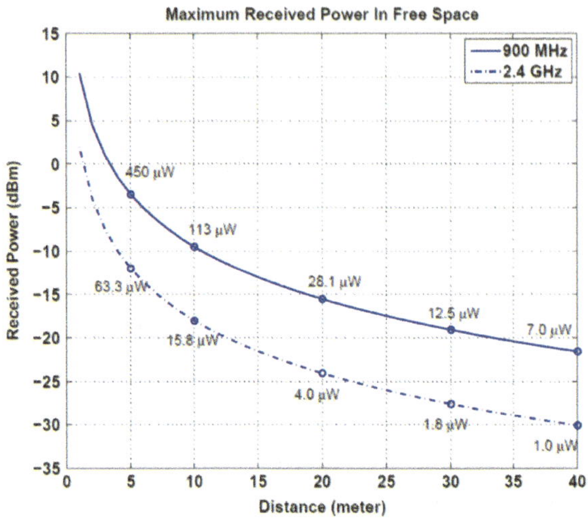

Fig. 2.24 Measurement of received power with the distance [59]

2.5.1 *Sensitivity*

It is the minimum incident power required for the activation of the circuit. Sensitivity helps the researchers to determine frequency range or the separation distance and make adjustments based on the application requirements. Sometimes, impedance matching or rectifying circuit affects the sensitivity. The MOSFET reduces the overall efficiency as they have more leakage current, as compared to other transistors which increase sensitivity.

2.5.2 *Resonator Q-factor*

The Q-factor characterizes the effect of resonance and impedance bandwidth. The voltage increases when a circuit resonates at its operating frequency. Theoretically, it is defined as the ratio of cut-off frequency and resonance bandwidth.

$$Q = 2\pi \frac{\text{Energy stored}}{\text{Energy dissipated per cycle}} = \frac{f_c}{\Delta f} \tag{2.17}$$

A short bandwidth results in a high Q-factor. In addition, the energy dissipation is defined in the form of reactive components. The Q-factor of the capacitor and inductor is defined as

$$Q_c = \frac{1}{\omega R_c C} = \frac{X_c}{R_c} \tag{2.18}$$

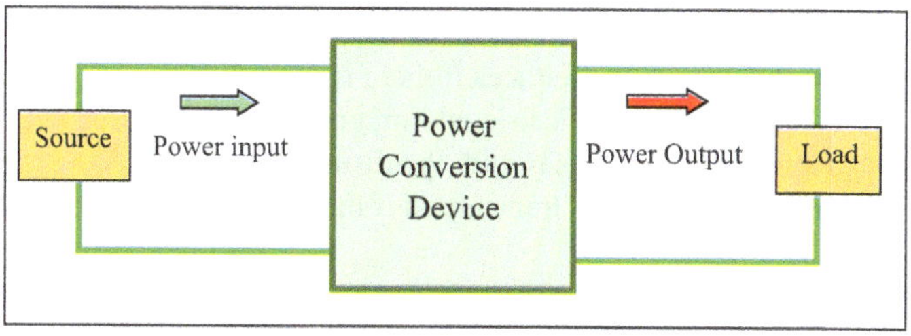

Fig. 2.25 Block diagram of the power conversion device

$$Q_L = \frac{\omega L}{R_L} = \frac{X_L}{R_L} \tag{2.19}$$

where L and C are inductance and capacitance, and R_L and R_C are series resistance of the inductor and capacitor, respectively. A high Q-factor is required for WPT. The loss and leakage currents can be minimized by reducing the resistive effect from the circuit and adding reactive components like capacitors and inductors.

2.5.3 *Power Conversion Efficiency*

This parameter characterizes the power conversion capability of any electrical circuit as shown in Fig. 2.25. It is simply the ratio of output power and input power in watts. It is represented by the Greek letter (η).

$$\eta = \frac{\text{Output Power}}{\text{Input Power}} \times 100 \tag{2.20}$$

Ideally, the transmission conversion efficiency should be 100%, but in practice, it is not possible to achieve 100% efficiency. The constraints that a researcher should keep in mind before designing a system are:

- **Money**: In a less efficient system, more and more power is transmitted to increase the conversion efficiency. This factor increases the amount of consumed energy, which affects the cost of the system.
- **Time**: The devices take more time to recharge if the output power is less.

2.5.4 *Operation Range*

The focus is to maximize the power conversion efficiency and coverage area with low input power density. The separation distance between the transmitter and receiver depends upon some parameters:

- **Free-space path loss**: This characterizes the loss of transmitted power in free space, and the distance range reduces if there is more loss.
- **EM waves**: The behavior of electric and magnetic fields decides the separation distance as their phase also varies with the distance.
- **Power leakage**: Less power is transmitted if there is power loss in the system.

References

1. Garnica J, Chinga RA, Lin J (2013) Wireless power transmission: from far field to near field. Proc IEEE 101(6):1321–1331
2. Valenta CR, Durgin GD (2014) Harvesting wireless power: survey of energy-harvester conversion efficiency in far-field, wireless power transfer systems. IEEE Microwave Mag 15(4):108–120
3. Michaelides EE (2012) Alternative energy sources. Green energy and technology. Springer, Berlin
4. Maxwell JC (1865) VIII. A dynamical theory of the electromagnetic field. Philos Trans R Soc (155):459–512
5. Hertz H (1970) Dictionary of scientific biography, vol VI. Scribner, New York, pp 340–349
6. O'Neill JJ (1944) Prodigal genius—the life of Nikola Tesla. Washburn, New York
7. Tesla N (1904) The transmission of electric energy without wires (The thirteenth anniversary number of the electrical world and engineer). McGraw-Hill, New York
8. Tesla N (2007) Nikola Tesla: Colorado spring notes, 1899–1900, 1st edn. BN Publishing
9. Brown WC (1964) Experiments in the transportation of energy by microwave beam. IEEE Int Conv Rec 12:8–17
10. Brown WC (1968) Thermionic diode rectifier. In: Microwave power engineering, vol I. Academic, New York, pp 295–298
11. Brown WC (1968) The combination receiving antenna and rectifier. In: Microwave power engineering, vol II. Academic, New York, pp 273–275
12. Brown WC (1969) Experiments involving a microwave beam to power and position a helicopter. IEEE Trans Aerosp Electron Syst AES-5(5):692–702
13. George RH, Sabbagh EM (1963) An efficient means of converting microwave energy to dc using semiconductor diodes. IEEE Intern Conv Rec Electron Devices Microwave Theory Tech 11:132–141
14. Brown WC (1984) The history of power transmission by radio waves. IEEE Trans Microwave Theory Tech MTT-32:1230–1242
15. Dickinson RM, Brown WC (1975) Radiated microwave power transmission system efficiency measurements. Jet Propulsion Laboratory, Calif Inst Tech 33:38
16. Glaser PE (1968) Power from the sun; its future. Science 162(3856):857–886
17. Glaser PE, Maynard OE, Macfcovciak J, Ralph EL (1974) Feasibility study of a satellite solar power station. NASA CR-2357, NTIS N74–17784
18. Matsumoto H, Kaya N, Fujita M, Fujino Y, Fujiwara T, Sato T (1993) MILAX airplane experiment and model airplane (in Japanese). In: Proceedings of 11th ISAS space energy symposium, pp 47–52
19. Fujino Y, Itoh T, Fujita M, Kaya N, Matsumoto H, Kawabata K, Sawada H, Onodera T (1993) A rectenna for MILAX. In: Proceedings of wireless power transmission conference, pp 273–277
20. Brown WC (1986) A microwave powered, long duration, high altitudeplatform. In: IEEE MTT-S international microwave symposium digest, vol 86, no 1, pp 507–510
21. Itoh K (1984) Study of rectenna as ground site of solar power satellite (in Japanese). Tech Rep Grant-in-Aid Scientific Res [Grant-in Aid Sci Res (A)]

22. Mcspadden JO, Mankins JC (2002) Space solar power programs and microwave wireless power transmission technology. IEEE Microwave Mag 46–57
23. Rozgic D, Markovic D (2017) A miniaturized 0.78-mW/cm^2 autonomous thermoelectric energy-harvesting platform for biomedical sensors. IEEE Trans Biomed Circuits Syst 11(4):773–783
24. Kamalinejad P, Mahapatra C, Sheng Z, Mirabbasi S, Leung VCM, Guan YL (2015) Wireless energy harvesting for the internet of things. IEEE Commun Mag 53(6):102–108
25. Karalis A, Joannopoulos JD, Soljačić M (2008) Efficient wireless non-radiative mid-range energy transfer. Ann Phys 323(1):34–48
26. Song C, Huang Y, Zhou J, Zhang J, Yuan S, Carter P (2015) A High-efficiency broadband rectenna for ambient wireless energy harvesting. IEEE Trans Antennas Propag 63(8):3486–3495. https://doi.org/10.1109/TAP.2015.2431719
27. Singh N, Kanaujia BK, Beg MT, Khan T, Kumar S (2018) A dual polarized multiband rectenna for RF energy harvesting. AEU-Int J Electron Commun 93:123–131
28. Singh N (2020) A compact and efficient graphene FET based RF energy harvester for green communication. Int J Electron Commun (AEÜ) 115:153059–153066
29. Yo TC, Lee CM, Hsu CM, Luo CH (2008) Compact circularly polarized rectenna with unbalanced circular slots. IEEE Trans Antennas Propag 56:882–886
30. Huang FJ, Yo TC, Lee CM, Luo CH (2012) Design of circular polarization antenna with harmonic suppression for rectenna application. IEEE Antennas Wireless Propag Lett 11:592–595
31. Zainol N, Zakaria Z, Abu M, Yunus MM (2018) A 2.45 GHz harmonic suppression rectangular patch antenna with circular polarization for wireless power transfer application. IETE J Res 64(3):310–316
32. Nie MJ, Yang XX, Tan GN, Han B (2015) A compact 2.45-GHz broadband rectenna using grounded coplanar waveguide. IEEE Antennas Wireless Propag Lett 14:986–989
33. Agrawal S, Gupta RD, Parihar MS, Kondekar PN (2017) Wideband high gain dielectric resonator antenna for RF energy harvesting application. AEU Int J Electron C 78:24–31
34. Shi Y, Jing J, Fan Y, Yang L, Pang J, Wang M (2018) Efficient RF energy harvest with a novel broadband Vivaldi rectenna. Microw Opt Technol Lett 60:2420–2425
35. Huang FJ, Lee CM, Chang CL, Chen LK, Yo TC, Luo CH (2011) Rectenna application of miniaturized implantable antenna design for triple-band biotelemetry communication. IEEE Trans Antennas Propag 59:2646–2653
36. Agrawal S, Parihar MS, Kondekar PN (2018) A dual-band rectenna using broadband DRA loaded with slot. Int J Microw Wirel Technol 59–66
37. Hassan N et al (2019) Design of dual-band microstrip patch antenna with right-angle triangular aperture slot for energy transfer application. Int J RF Microw Comput Aided Eng 29(1):1–11
38. Singh N, Kanaujia BK, Beg MT, Siddique M, Kumar S, Khandelwal MK (2018) A dual band rectifying antenna for RF energy harvesting. J Comput Electron 17:1748–1755
39. Chandravanshi S, Sarma SS, Akhtar MJ (2018) Design of triple band differential rectenna for RF energy harvesting. IEEE Trans Antennas Propag 66(6):2716–2726
40. Noor FSM, Zakaria Z, Lago H, Said MAM (2019) Dual-band aperture-coupled rectenna for radio frequency energy harvesting. Int J RF Microw Comput Aided Eng 29(1):1–9
41. Kumar H, Arrawatia M, Kumar G (2017) Broadband planar log-periodic dipole array antenna based RF-energy harvesting system. IETE J Res 65(1):39–43
42. Nguyen NH et al (2018) A novel wideband circularly polarized antenna for RF energy harvesting in wireless sensor nodes. Int J Antennas Propag 2018:1–9
43. Shen S, Chiu CY, Murch RD (2018) Multiport pixel rectenna for ambient RF energy harvesting. IEEE Trans Antennas Propag 66(2):644–656
44. Adam I, Yasin MNM, Rahim HA, Soh PJ, Abdulmalek MF (2018) A compact dual-band rectenna for ambient RF energy harvesting. Microw Opt Technol Lett 60(11):1–9
45. Shinohara N, Matsumoto H (1998) Dependence of dc output of a rectenna array on the method of interconnection of its array elements. Electr Eng Jpn 125(1):9–17

46. Joshi S, Moddel G (2016) Simple figure of merit for diodes in optical rectennas. IEEE J Photovoltaics 6(3):668–672. https://doi.org/10.1109/JPHOTOV.2016.2541460
47. Wang K, Hu H, Lu S, Guo L, He T (2015) Design of a sector bowtie nano-rectenna for optical power and infrared detection. Frontiers Phys 10(5):104101
48. Grover S, Joshi S, Moddel G (2013) Quantum theory of operation for rectenna solar cells. J Phys Appl Phys 46(13)
49. Arrawatia M, Maryam B, Kumar G (2011) RF energy harvesting system from cell towers in 900 MHz band. In: National conference on communications, pp 1–5
50. Department of Telecommunications, Ministry of Communications and Information Technology, Annual Report, 2012–2013
51. Pan J, Gao X, Fan J. Identifying Interference from multiple noise sources using only magnetic near fields. IEEE Trans Electromagn Compat. https://doi.org/10.1109/TEMC.2020.3020049
52. Puniran KN, Ahmad R, Dziyauddin RA (2017) RF energy harvesting with multiple sources in wireless sensor network. In: 2017 IEEE 4th international conference on smart instrumentation, measurement and application (ICSIMA), Putrajaya, pp 1–5. https://doi.org/10.1109/ICSIMA.2017.8312005
53. Lumpkins W (2014) Nikola Tesla's dream realized: wireless power energy harvesting. IEEE Consum Electron Mag 3(1):39–42. https://doi.org/10.1109/MCE.2013.2284940
54. Zhong C, Suraweera HA, Zheng G, Krikidis I, Zhang Z (2014) Wireless information and power transfer with full duplex relaying. IEEE Trans Commun 62(10):3447–3461. https://doi.org/10.1109/TCOMM.2014.2357423
55. Hirayama H, Ozawa T, Hiraiwa Y, Kikuma N, Sakakibara K (2009) A consideration of electromagnetic resonant coupling mode in wireless power transmission. IEICE Electron Express 6(19):1421–1425
56. Zhang J, Huang Y, Cao P (2014) A microwave wireless energy harvesting system with a wideband antenna array. Trans Inst Meas Control 37(8):961–969
57. Hong H, Cai X, Shi X, Zhu X (2012) Demonstration of a highly efficient RF energy harvester for Wi-Fi signals. In: Microwave and millimeter wave technology (ICMMT) 2012 international conference, vol 5, pp 1–4, May 2012
58. Uzun Y (2016) Design and implementation of RF energy harvesting system for low-power electronic devices. J Electron Mater 1–6
59. Alippi C, Galperti C (2008) An adaptive system for optimal solar energy harvesting in wireless sensor network nodes. IEEE Trans Circuits Syst 55(6):1742–1750
60. Pareja Aparicio M, Bakkali A, Pelegri-Sebastia J, Sogorb T, Llario V, Bou A (2016) Radio frequency energy harvesting—sources and techniques. In: Renewable energy—utilisation and system integration. InTech

Chapter 3
Antennas

3.1 Introduction

Antenna is a key element of the wireless communication system. It acts as a transducer and transforms electrical energy into electromagnetic waves and vice versa. With the increasing demand for antennas for contemporary wireless applications, continuous improvement has been seen in antenna technologies. Printed antennas are lightweight, low-profile, easy to manufacture, and possess high-performance characteristics. Since printed antennas are compact, flexible, and durable with low manufacturing cost, they are a potential candidate for aerospace, military, medical, and commercial wireless devices. Moreover, the easy integration of the patch antenna with the circuitry of wireless systems extends their area of application, such as smart imaging and sensing, power transfer, energy harvesting, and 5G phased array.

The operational frequency of the printed antenna ranges from kilohertz to millimeter waves. Printed antennas are utilized in classical microwave applications, such as aeronautical, radar, biotelemetry, space, mobile and wireless communication, and global positioning systems (GPS). They are often used in smart cards, point-of-sale machines, product packaging, e-textiles, collection of toll electronically, and equipment tracking and identification. Printed antenna generally possesses a resonator characteristic and thus provides a narrow bandwidth. However, various technologies have been reported for changing the antenna performance and make it function for broadband applications.

3.2 Types of Printed Antennas

Based on the feed or excitation port, radiating element, and radiation mechanism [1–5], the printed antennas are classified as follows.

B. K. Kanaujia et al., *Rectenna: Wireless Energy Harvesting System*,
Advances in Sustainability Science and Technology,
https://doi.org/10.1007/978-981-16-2536-7_3

3.2.1 Microstrip Antenna

A microstrip antenna consists of a patch, a ground plane, and the dielectric substrate. The dielectric substrate separates the radiating patch and the ground plane. The researchers proposed different shapes of the radiating patch; however, circular and rectangular shapes are most common [6]. The schematic of the simple microstrip patch antenna is shown in Fig. 3.1. The patch and the ground plane are printed on the top and bottom surfaces of the dielectric substrate, respectively [7]. The relative permittivity (ε_r) of the dielectric substrate usually varies from 2.2 to 13. The lower permittivity substrates are used to obtain larger bandwidth and high efficiency, whereas higher permittivity substrates ($\varepsilon_r > 10$) are used to obtain compact designs. The half-wave rectangular-shaped patch has simple geometry and is often used [8, 9].

A customary microstrip antenna is large in size, but the technological breakthrough in the miniaturization of the antenna design resulted in a small size with easy integration into the electronic circuitry. The radiating patch of the microstrip antenna can be printed using various conducting materials such as gold or copper. Few shapes of the patch, such as circular, dipole, rectangular, or square, are commonly used as they provide better radiation patterns and easy fabrication. Different types of feed such as proximity coupling, aperture coupling, coaxial probe, and microstrip line can be used to excite the patch antenna.

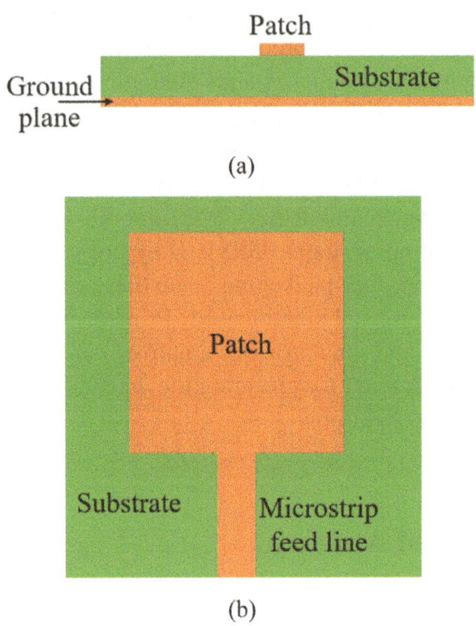

Fig. 3.1 Simple microstrip patch antenna: **a** side view, **b** top view

3.2.2 Printed Dipole Antenna

A dipole antenna consists of two radiating arms located on different sides of the dielectric substrate. If the two radiating arms are fed in opposite phases, they may be located on the same side also. Various types of feeding techniques are used for exciting the antenna depending on the position of the arms. Coaxial and coplanar waveguide (CPW) line feeds are commonly used for exciting the dipole antenna. Different types of arm shapes can be used as the dipole. Printed dipole antennas are widely used in personal wireless systems such as Wi-Fi, Wi-MAX, and Bluetooth.

3.2.3 Monopole Antenna

Monopole antennas consist of only one radiating arm that is fed directly using a transmission line. Monopole antennas are used for wideband applications in mobile and wireless devices. As compared to the dipole antennas, the monopole antennas are compact and offer a broad bandwidth.

3.2.4 Slot Antenna

A slot antenna is formed by removing pertinent metal from the metallic surface of the substrate [10, 11]. Slot antennas have wider bandwidth as compared to the microstrip patch antennas. The slot antenna shows a bidirectional radiation pattern, and a metallic ground plane can be employed to achieve unidirectional radiation patterns. Figure 3.2 displays a schematic of the slot antenna. Ring-shaped and rectangular-shaped slot antennas are most often used. The integration of the slot antenna with the circuitry of the wireless electronic devices is difficult; therefore, a CPW feed is typically used to simplify the integration process. Substrate integrated waveguide (SIW) integrated slot antennas have recently attracted a lot of attention

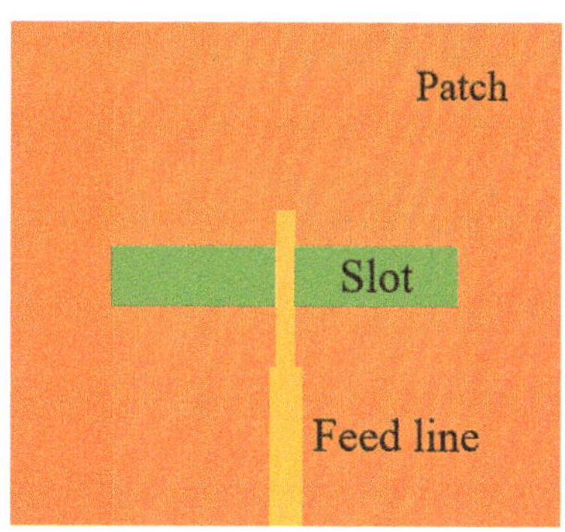

Fig. 3.2 Slot antenna structure

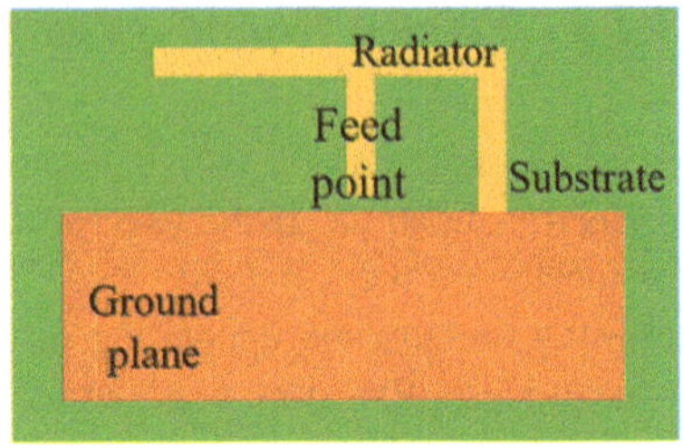

Fig. 3.3 Inverted-F antenna configuration

from researchers. In comparison to the conventional slotted antennas, the slotted SIW antennas are more advantageous as they are compact and can be simply integrated with other wireless devices due to their waveguide behavior.

3.2.5 Inverted-F Antenna

The inverted-F antennas (IFAs) are widely employed in phones and laptops [12–14]. The layout of the IFA is shown in Fig. 3.3. The inverted-F is a monopole configuration, where the antenna size can be reduced by using the meandered trace, and two or more F-structures can be stacked together to achieve multiband operation. Other antenna designs, such as a monopole, inverted-L, and loops, may also be used with IFA. IFAs are designed in such a manner that they can easily fit within the product geometry or the available space.

3.2.6 Planar Inverted-F Antenna

The planar inverted-F antenna (PIFA) is a category of microstrip antenna and is different from the IFA. PIFA is similar to the patch antenna, whereas IFA is similar to the monopole [15–17]. The planar shape of the PIFA allows more volume in the device and it has a larger bandwidth than the IFA. In PIFA, a compact size radiating patch is printed on the large-sized ground plane, and a metallic pin or a metallic surface is used to connect the ground plane and the patch, shown in Fig. 3.4.

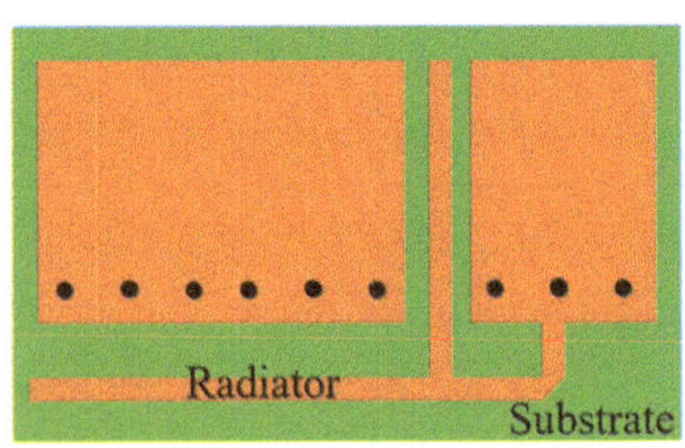

Fig. 3.4 Planar inverted-F antenna configuration

Fig. 3.5 Printed inductor antenna configuration

The PIFAs are widely used in laptops and mobile phones due to their multiband behavior. The PIFA allows very little radiation to pass through the ground surface, and therefore, less specific absorption rate (SAR) is noticed in them.

PIFA comes with different shapes and designs of the radiating patch. These antennas are characterized by limited space with low cost, diversity, robustness, lightweight, multiband, packaging capability, and integration of MEMS/RF PIN switches for smart systems. Various techniques have been tried to reduce the size of the antenna such as dielectric loading, use of shorting plates, etc. The dielectric loading technique is generally not used as it increases the fabrication cost and decreases the bandwidth.

3.2.7 Printed Inductor Antenna

Inductor antennas are electrically small antennas with inductive coupling and work in the near-field region. Their specific frequency varies according to their application [18]. The inductor antenna, which works as the main radiating element in the RFID tag, is widely used at UHF-band (900 MHz). A schematic of the printed inductor antenna for RFID application is shown in Fig. 3.5.

3.2.8 Printed Quasi-Yagi-Uda Antenna

A printed quasi-Yagi antenna consists of a Yagi-Uda antenna and an additional parasitic element. Over the wired Yagi-Uda antenna, the planar quasi-Yagi antenna has a low profile, lightweight, simple structure, high gain, broad bandwidth, low

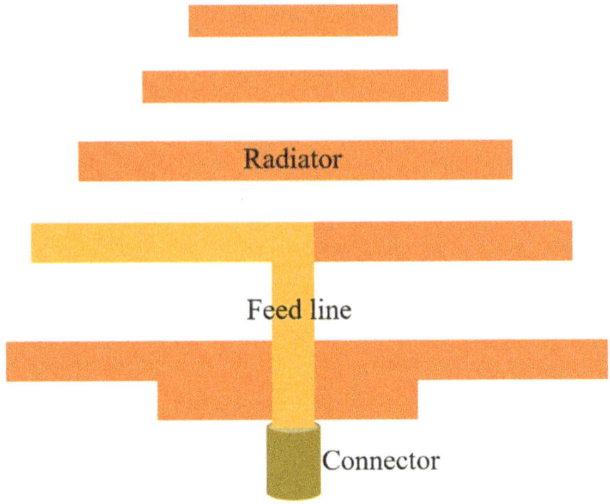

Fig. 3.6 Printed quasi-Yagi-Uda antenna structure

fabrication cost, unidirectional radiation, and it can be easily integrated with high-frequency microwave circuits. Planar Yagi-Uda antennas can be used in wireless sensor networks (WSNs) to increase the communication range [19].

The quasi-Yagi-Uda antenna, despite having a narrow bandwidth, is used for multiple wireless applications due to its high gain and end-fire radiation pattern. A schematic of the printed quasi-Yagi-Uda is illustrated in Fig. 3.6.

3.2.9 Log-Periodic Antenna

Log-periodic antennas possess frequency-independent behavior and are used to achieve wider bandwidth. A log-periodic dipole array (shown in Fig. 3.7) is an example of a log-periodic antenna. It consists of complementary radiating elements and is driven by half-wave dipoles. Large size is the main disadvantage of the log-periodic antennas [20, 21].

3.2.10 Fractal Antenna

Benoit Mandelbrot used the fractal term for the classification of structures, whose dimensions were not whole numbers. Fractals are marked by the property of infinite length within a finite volume. In the available literature, various fractal shapes, for example, Minkowski and Koch, have been used to obtain a wide bandwidth and significant reduction in the size of the antenna. The Sierpinski fractal shape was used for multiband applications [22–25] (Fig. 3.8).

Fig. 3.7 Log-periodic antenna design

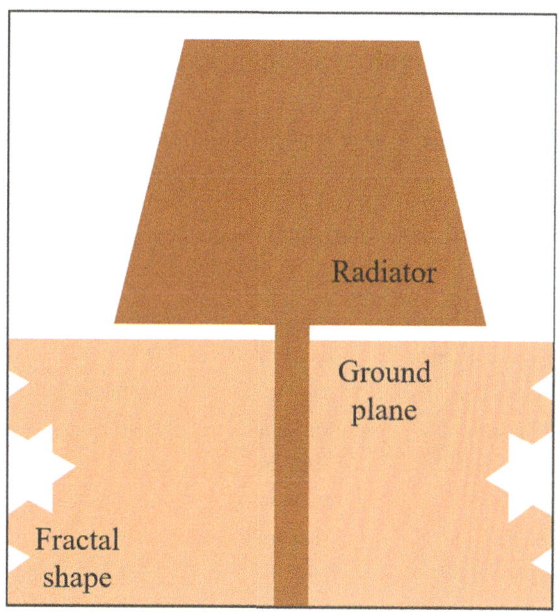

Fig. 3.8 Layout of the fractal antenna

3.2.11 Customized Printed Antenna

A printed antenna is designed according to the available space and the shape of the electronic device, where it needs to be positioned. Customized printed antennas can be easily integrated into different wireless communication devices to provide high throughput and data transmission rate. The application of the customized printed antenna, in fields such as mobile communication, defense, and aerospace, is increasing day by day. Customized printing also allows the integration of many antennas in a single device.

3.2.12 Comparison of the Planar Antennas

A comparison of different forms of planar antennas is presented in Table 3.1. Slot and patch antennas are compact and low profile, which can be easily incorporated within a large-size system. IFA, PIFA, and customized antennas cover multiple standards and are mostly used for mobile communication. Log-periodic and quasi-Yagi-Uda antennas show the ability of frequency scanning and possess better directivity.

3.3 Important Specifications of Antenna Design

The parameters necessary to understand the antenna design are explained in this section [1–5]. The antenna design parameters include voltage standing-wave ratio (VSWR), polarization, side lobe, half-power beamwidth (HPBW), return loss, impedance, bandwidth, gain, frequency, and cross-polarization.

Table 3.1 Comparison of the planar antenna designs

Type of antenna	Bandwidth	Directivity	Radiation pattern
Microstrip	Narrow	Medium	Broadside
Monopole	Medium	Low	Broadside
Dipole	Medium	Low	Broadside
Slot	Medium	Low/medium	Broadside/bidirectional
PIFA	Medium	Medium	Broadside
Quasi-Yagi-Uda	Wide	High	End-fire
Log-periodic	Wide	Medium	End-fire
Fractal	Wide	High	Broadside

3.3.1 Working Frequency

The antenna design frequency (f_c) or the resonant point is the frequency at which the antenna resonates. The operating range of the antenna is known as impedance bandwidth, and it corresponds to a range, where the antenna performance follows a specific standard with respect to some characteristics. The center frequency of the antenna lies between a low-frequency point (f_L) and a high-frequency point (f_H) and can be calculated using the Eq. (3.1).

$$f_c = \frac{f_H - f_L}{2} \tag{3.1}$$

3.3.2 Impedance

The impedance provides a relation between the input current and the voltage. It is made up of two components: real resistive value and imaginary reactance value. The capacitive and inductive components vary with frequency. The two resistance types associated with an antenna are ohmic resistance and radiation resistance. Ohmic resistance changes the electrical power into heat while the radiation resistance changes it into radiation. The value of radiation resistance must be greater than the ohmic resistance to obtain higher efficiency. The impedance of the antenna system and the transmission line should match for the efficient transfer of energy (Fig. 3.9).

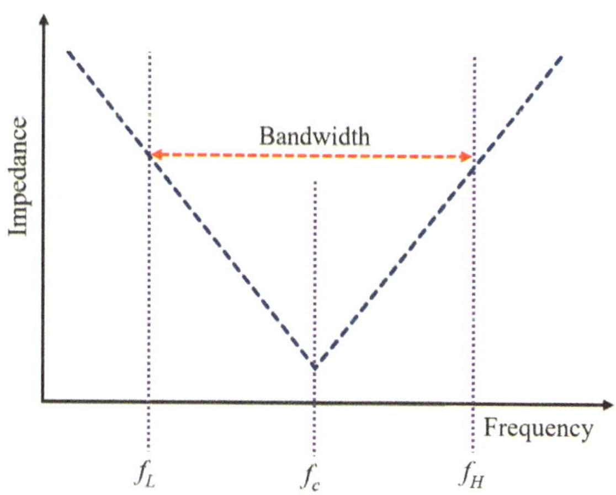

Fig. 3.9 Relation between impedance and bandwidth

3.3.3 Return Loss and VSWR

The return loss signifies the power delivery efficiency of the antenna. For an antenna under test, if the incident power is P_{in} and the reflected power is P_{ref}, then the ratio P_{in}/P_{ref} shows the degree of mismatch between the two powers. Return loss is expressed either by the returned signal levels or by the VSWR. It is measured in decibel (dB). The return loss can be calculated by the following Eq. (3.2).

$$\mathrm{RL(dB)} = 10\log_{10}\frac{P_{in}}{P_{ref}} \tag{3.2}$$

VSWR and return loss are related to each other. VSWR signifies the amount of power reflected from the source. An ideal VSWR signifies the ratio of 1:1. The VSWR value is usually 2:1, which illustrates that the antenna radiates 88.9% of the incident energy and 11.1% of the incident energy is either lost in the form of heat during transmission or is reflected back into the source. Return loss is also used as the statistics to measure the bandwidth of the antenna.

3.3.4 Radiation Pattern

Radiation pattern is defined as the variations of the radiated power at different locations in space as a function of the direction. The measurement of radiated power is performed in the far-field of the antenna. The isotropic antenna pattern signifies that the radiation patterns are the same at all angles. An isotropic antenna does not exist in practice, but it is used as the reference for the comparison of the gain and the radiation patterns of the real antennas.

The 2-D radiation patterns are plotted for different elevation and azimuth angles, which are known as elevation and azimuth plane patterns, respectively. The radiation pattern can be plotted in the rectangular or the polar form. The rectangular plot provides more information if the side lobes are present. Various parameters, such as null locations, main lobe, side lobes, HPBW, directivity, and gain, can be calculated using the radiation patterns of the antenna as shown in Fig. 3.10.

3.3.5 Directivity and Gain

Directivity is the measure of the antenna's ability to focus the radiated energy in a direction. The directivity quantifies the amount of radiation density (in a particular direction) with reference to a radiating element while the same source power is applied in both cases. The isotropic antenna or half-wave dipole antenna is used as the reference antenna for measuring the directivity. Theoretically, the isotropic

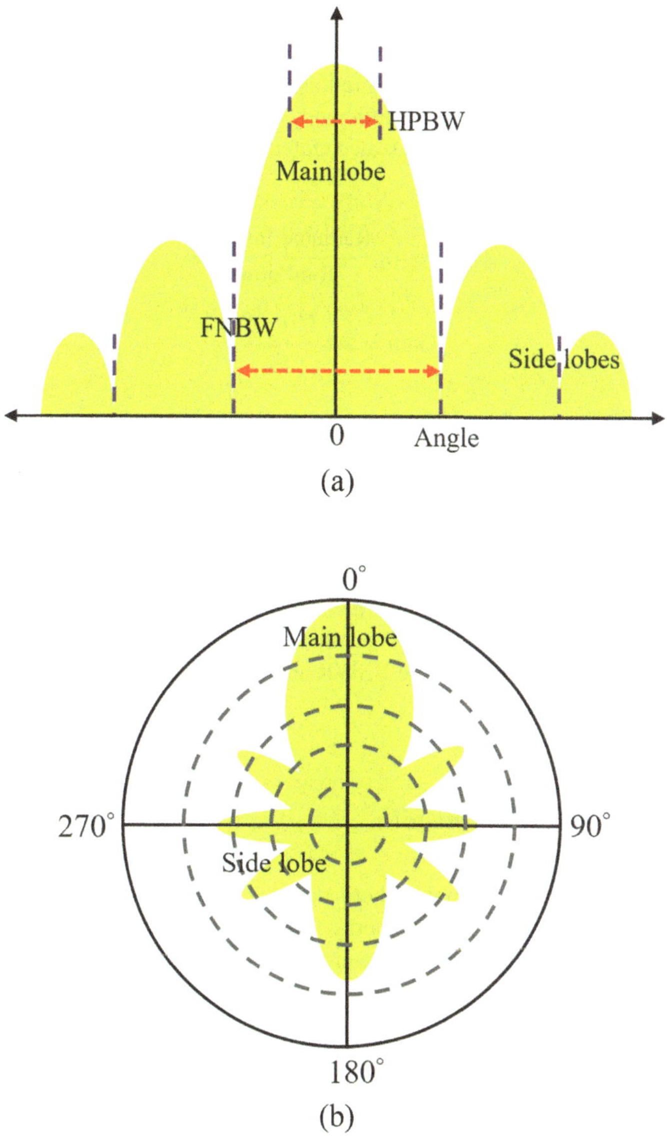

Fig. 3.10 Radiation patterns of the antenna: **a** rectangular plot, **b** polar plot

antenna shows equal radiation patterns in every direction, and thus, it is used as a reference for comparing the performance of the real antenna. The unit used for measuring the gain is decibels or dBd (with reference to the dipole antenna) or dBi (with reference to the isotropic antenna). The gain of the dipole antenna encompasses the gain of the isotropic antenna by 2.15 dB and can be calculated using the following relation.

$$\text{dBi} = \text{dBd} + 2.15\ \text{dB} \tag{3.3}$$

The amount of incident energy being radiated into the free space is used to measure the performance of the antenna. The efficiency of the antenna is always below 100% due to internal losses. These losses are accounted to calculate the gain of the antenna, which can be expressed as

$$\text{Gain} = 4\pi \frac{\text{Radiated intensity}}{\text{Total power}} \tag{3.4}$$

$$\text{Gain} = 4\pi \frac{U(\theta, \emptyset)}{P_{\text{in}}} \tag{3.5}$$

3.3.6 Antenna Efficiency

The energy radiated by the antenna into the free space is known as antenna efficiency. Generally, a lot of input power is returned back to the source or lost in the form of heat on the surface of the antenna. The antenna is efficient if it radiates most of the input power into the free space. The VSWR and the antenna impedance are used for the measurement of efficiency.

3.3.7 Half-Power Beamwidth

Beamwidth is the aperture angle where most of the energy is radiated from the antenna. The angle over which the energy falls to half of its peak level is known as HPBW or the − 3 dB point and is measured in degrees or radians.

3.3.8 Side Lobes

In an antenna, most of the power is radiated in the main beam and much lesser power density is present in the side lobes. Side lobe levels are measured in dB with reference to the main beam. Based on the application, the levels of the side lobes can be controlled using various techniques.

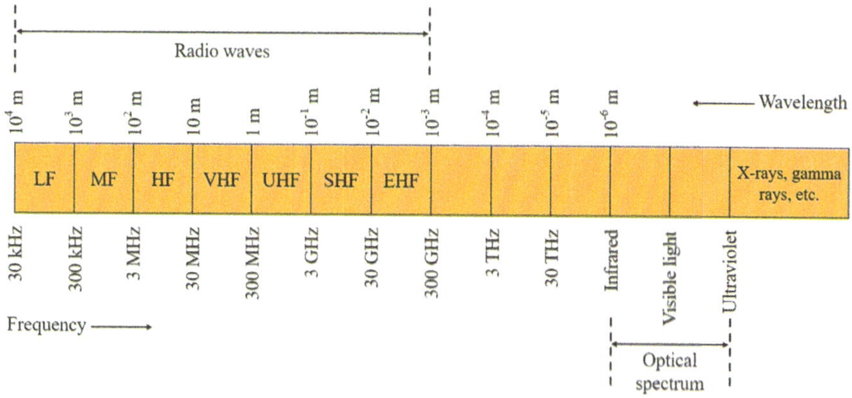

Fig. 3.11 RF spectrum

3.3.9 Polarization

Polarization is defined as the orientation of the electric field or the E-plane. The polarization can be elliptical, circular, or linear. Linear polarization is usually vertical or horizontal. The circularly polarized wave can be right-handed or left-handed polarized depending on the direction of the rotation of the wave. Antennas are never perfectly polarized. Cross-polarization is used to measure energy in the perpendicular plane with reference to the peak gain.

3.4 RF/Microwave Frequency Bands

The microwave range is separated into multiple bands as displayed in Fig. 3.11. The radio frequency band ranges from 1 to 300 GHz. These bands are assigned by different organizations like IEEE, NATO, and ITU. The different bands, their frequencies, and their applications are mentioned in Table 3.2.

3.5 Energy Harvesting

The conversion of energy available in the surroundings into electrical energy is known as energy harvesting. The energy harvested from various sources, such as light, heat, variations, radio frequency, etc., could serve as an alternate source of energy. It can also be used to charge the battery of wireless devices for remote sensing, wireless computing, and mobile phones [26, 27]. A comparison of various energy harvesting sources is presented in Table 3.3. Energy harvesting is advantageous in systems or

Table 3.2 RF/microwave frequency spectrum

Band	Frequency	Application
Low frequency (LF)	30–300 kHz	Amateur radio, navigation, submarine communication
Medium frequency (MF)	300 kHz–3 MHz	Amateur radio, AM transmission, marine navigation
High frequency (HF)	3–30 MHz	Mobile, RFID, shortwave broadcast
Very high frequency (VHF)	30–300 MHz	Amateur radio, FM, land mobile and maritime mobile communications, television broadcast
Ultra-high frequency (UHF)	300 MHz–3 GHz	Microwave communication, mobile phone, radio astronomy, satellite radio, wireless LAN
Super high frequency (SHF)	3–30 GHz	Microwave devices, radar communication, satellite television broadcasting
Extremely high frequency (EHF)	30–300 GHz	Millimeter-wave scanning, radio astronomy, remote sensing
Terahertz (THz)	30–300 GHz	Computing/communication, medical imaging, remote sensing, time-domain spectroscopy

Table 3.3 Comparison of various energy harvesting sources

Source	Energy type	Power density	Advantages	Limitations
Solar	Radiant	100 mW/cm^3	Available in large quantity	Does not available in the night, low efficiency
RF	Radiant	40 $\mu W/cm^2$ at 10 m	Available in large quantity	Low efficiency
Heat	Thermal	135 $\mu W/cm^2$ at 10 °C	Easy to use by means of thermocouple	Available only when temperature difference is high
Motion	Mechanical	800 $\mu W/cm^2$	High power density	Available only during the motion
Wind	Mechanical	177 $\mu W/cm^3$	High power density	Low conversion efficiency
Vibration	Mechanical	4 $\mu W/cm^3$	High power density	Low conversion efficiency
Piezoelectric	Mechanical	50 $\mu W/cm^2$	High power density	Does not exist at surrounding

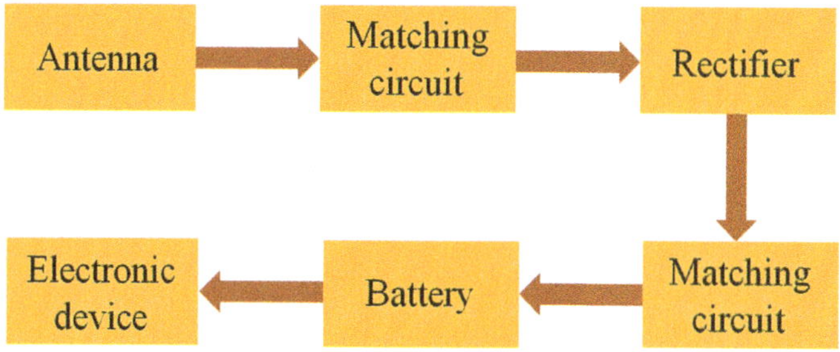

Fig. 3.12 Block diagram of RF energy harvesting system

machines used in the hazardous chemical industry where limited human access is possible.

A rectenna can be used to harvest ambient energy from radio frequency signals. The received power can be increased by using the antennas with wide bandwidth and high gain. Various energy harvesting antenna designs were reported in the literature [28–30]. In [29], an antenna with ~20 MHz bandwidth and 11.98 dB gain was reported. The matching factor between the antenna and the rectifier affects the overall efficiency of the system. The system efficiency also varies with the input power and frequency as they impact the input impedance of the rectifier. As the semiconductor technology is moving toward low power operation, the battery-operated devices could be prepared to use some other alternative power source, such as DC power generators, which can harvest energy from the environment. Various devices such as wireless LAN, mobile phones, base stations, radar, radio broadcasts, and TV can be used for RF energy harvesting applications. The harvested DC power can be used directly to power an electronic device or it can be stored in the capacitor for the operation of electrical devices. Figure 3.12 shows the block diagram of a simple energy harvesting system.

3.6 RF Energy Harvesting

The increasing power demand of the wireless devices can be met with the ambient RF energy. The RF energy is available throughout the day and night. The RF energy harvesting comes with the advantage of 24 h availability, quasi-independence from weather conditions, and no physical effort is needed for charging, and, therefore, it seems to be a promising choice for consumer electronic equipment. A charging setup based on ambient energy harvesting is very much useful when the wireless battery-operated device is located in an inaccessible place. The DC power, generated by rectifying the microwave signals, was studied to power solar power satellites and helicopters [31–33].

In the Netherlands, a study was conducted on the GSM power present in the surroundings, and it was observed that the average power present within the distance of 25–100 m from the base station was ~0.1 to 3.0 mW/m^2. The ambient average power varies highly with the GSM traffic present at the time of taking the measurements. The availability of the variable, unpredictable, and small quantity of energy poses a challenge for energy harvesting devices. Therefore, the energy harvesting device should be designed in such a way that it can operate over a range of input power.

The generated DC power depends on various factors, such as energy harvesting technique, the frequency band used, antenna, and available RF power. The factors considered for choosing energy sources for an application include magnitude, predictability, and controllability. The growing dependence on wireless devices such as GPS units, tablets, PCs, cellular phones, etc., for information gathering, scheduling, and daily navigation activities has created a demand for longer battery life [34].

RF energy harvesting finds its application in various areas such as aircraft fatigue supervision, long-range asset tracking, implantable sensors, home automation, agricultural management, tracking enemy troop movement, efficient office energy control, harmful agent detection, and remote patient monitoring. Billions of radio transmitters, including television/radio broadcast stations, mobile base stations, handheld radios, and mobile phones, radiate RF energy into the air. Rectenna can be used to efficiently harvest the ambient energy into DC current. The rectenna mainly comprises of a DC pass filter, a rectifying circuit, a pre-rectification filter, and an antenna. Figure 3.13 highlights the collection process of incoming RF energy by the antenna. A low-pass filter is used at the input of the circuit to suppress the unwanted higher-order harmonics which are then rejected by the rectifier circuit. It also participates in providing a match among the rectifier and the antenna [35]. A customary rectenna uses a dipole antenna for capturing the energy and Schottky diode to perform the rectification step. But, recently various types of rectenna designs have been proposed that employs loop, monopole, dipole, microstrip, Yagi-Uda, coplanar, parabolic, or even spiral antenna [36] for energy harvesting. A dual diode rectifier is used to enhance the conversion efficiency, whereas a half-wave rectifier is used to generate double-output DC voltage [37].

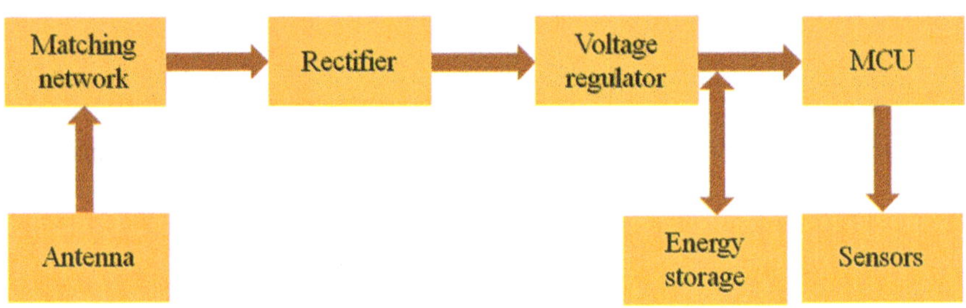

Fig. 3.13 Block diagram of a typical power harvesting circuit

3.7 Antenna Designs Used for RF Energy Harvesting

Various antenna designs, such as patch antenna, monopole antenna, dipole antenna, folded dipole, fractal antenna, PIFA, circular patch antenna, spiral antenna, gap-coupled microstrip antenna, slotted patch antenna, and arrays, have been used for energy harvesting applications. To work as an energy harvesting system, an antenna should offer [38]:

(i) Circular polarization
(ii) Multiband operation
(iii) Matching impedance to transfer the maximum amount of power to the following rectifier.

Different types of antennas, ranging from simple dipole to complex shapes like spiral or bowtie, have been used for RF energy harvesting. These antennas have better performance in terms of polarization, but are usually limited to broadband designs. The requirement of a complex feed arrangement, by each element of the antenna, limits the use of multiple narrowband frequency designs.

In circularly polarized antennas, the output voltage does not vary with the movement of the receiver or the transmitter, and therefore, it constitutes one of the important characteristics of the rectenna design [39]. Though the circularly polarized antennas are immune from the rotation of the transmitter/receiver, still, the output voltage varies if there is no proper alignment of the main beam among the receiving and the transmitting antennas. The output voltage can be kept constant, even when a proper main beam alignment is not present, by using a broad beamwidth rectenna.

3.8 Recent Trends in RF Energy Harvesting Antennas

In the literature, different antenna configurations have been presented for RF energy harvesting. In [40], the antenna was designed at 2.45 GHz, where two coupling slots were arranged orthogonally to obtain multi-polarization. A multi-layered rectenna consisted of a circular patch, concentric annular rings, phase shifter/rectifier, slotted ground plane was designed for 900, 1760, and 2450 MHz frequency bands [41]. In [42], an antenna with a central radiating element and two coplanar parasitic elements was designed. In [43], the sixty-four element array was implemented by connecting the spiral elements in series and parallel to obtain right-hand circularly polarized (RHCP) and left-hand circularly polarized (LHCP) signals.

PIFA is increasingly being used in mobile phones and other devices that operate in the ISM band. In [44], a PIFA antenna was proposed for energy harvesting in the 2.45 GHz band. As compared to the printed antenna, the design of the PIFA is complex due to the fact that the ground and the feed point in the PIFA are subject to fragility.

3.8.1 Transparent Antenna

The concept of the transparent antenna was given by the National Aeronautics and Space Agency (NASA). Different materials, such as aluminum-doped zinc oxide (AgHT) and indium tin oxide (ITO), can be used to develop transparent antennas. In addition to these materials, a coating of transparent polymer substrates with traces of non-transparent conductive ink can be used to construct such antennas [45].

The transparent antenna is mainly used to convert RF energy into DC power. The main challenge faced is to obtain high conversion efficiency. The different methods used to achieve high conversion efficiency are:

- The transparent antennas are designed on conductive, conformal, flexible, or discreet materials, which provide good performance on the glass.
- Use of AgHT and transparent conductive oxide (TCO) polymer for the antenna design, which can be integrated with photovoltaic and can be used on the glass of the windows in homes.

Besides this, the use of transparent antennas for on-body wireless communications has also been explored [46, 47]. One of the common uses of transparent conductive oxides is their ability to reflect thermal infrared heat, which can be used in the manufacturing of energy conserving windows. It collects and delivers a reasonable amount of power to the rectifying diode and also suppresses the diode generated harmonics reradiated from the antenna. Multiple broadband antennas, circularly polarized antennas, and large antenna arrays are used to increase the RF to DC conversion efficiency. The antenna array can be used to increase the received power, but it results in large antenna size. Different types of transparent antennas can be used as the component of the rectenna, but a meshing microstrip patch antenna is preferred as it is simple, low weight, low profile, low cost, and can be easily manufactured using printed circuit technology [48].

In a simple wireless sensor system, a transducer converts the surrounding RF energy into DC and stores it in a supercapacitor or a battery for further use. A power management unit performs matching and duty-cycle optimization to maximize the collected power in an efficient way. The energy harvesting system itself serves as a power source or it can be used to periodically recharge the power source, such as a battery to increase the overall lifetime of the power source. The substrate is made up of glass, while the ground and the radiating element are designed using a transparent conductive oxide polymer. The transparent antennas can also be used in integration with solar cells to provide reduced surface area.

3.8.2 Reconfigurable Antennas

Reconfigurable antennas are in increasing demand for RF energy harvesting applications as their polarization, tuning, and operating frequency can be changed. The different antenna parts that alter the radiation of the antenna can be connected and/or disconnected at different times to modify the physical structure of the antenna, and thus its RF response. A multiband antenna uses frequency diversity to allow the transmission of data, voice, and video without using two different antennas.

The wideband frequency is usually achieved using two different techniques: using stacked patches or by activating different staggered modes of the patch. The first technique involves the use of a multi-layered patch substrate resonating at different frequencies, but the antenna height should be kept small. The second technique involves the working of two frequencies, which is achieved by two different modes, such as TM_{01} and TM_{10} modes or TM_{10} and TM_{30} modes. Primarily, the elements of the multi-layered patch antenna radiate linearly polarized waves, but the circular and elliptical polarization can also be achieved by varying the shape of the elements and the position of the feed. The circular polarization can be obtained by exciting two orthogonal modes that are 90° out of phase with each other. A constant DC voltage is achieved in the circularly polarized rectenna, and it is not affected by the rotation of the rectenna, and thus overcomes the problem of polarization mismatch.

In reconfigurable antennas, different parts of the antenna can be connected or disconnected using the switching elements such as photoconductive switches, PIN diodes, RF micro-electro-mechanical systems, or lumped elements.

References

1. Waterhouse RB (2003) Microstrip patch antennas: a designer's guide. Kluwer Academic Publishers, Norwell, MA
2. Sangeetha RG (2015) Microstrip patch antenna array with omnidirectional pattern. Lambert Academic Publishing
3. Al-Naiemy Y (2013) Design, fabrication, and testing of microstrip antennas. Lambert Academic Publishing
4. Bhartia P, Bahl I, Garg R, Ittipiboon A (2001) Microstrip antenna design handbook. Artech House, Norwood, MA
5. Kumar G, Ray KP (2002) Broadband microstrip antennas. Artech House, Norwood, MA
6. Pan BC, Cui TJ (2017) Broadband decoupling network for dual-band microstrip patch antennas. IEEE Trans Antennas Propag 65(10):5595–5598
7. Mendes C, Peixeiro C (2017) A dual-mode single-band wearable microstrip antenna for body area networks. IEEE Antennas Wirel Propag Lett 16:3055–3058
8. Pandey BK, Pandey AK, Chakrabarty SB, Kulshrestha S, Solanki AL, Sharma SB (2006) Dual frequency single aperture microstrip patch antenna element for SAR applications. IETE Tech Rev 23(6):357–366
9. Pandey A (2013) Design of a cosecant square-shaped beam pattern SAR antenna array fed with square coaxial feeder network. In: 2013 European radar conference (EuRAD), Nuremberg, Germany, pp 1699–1702
10. Antoniades MA, Dadgarpour A, Razali AR, Abbosh A, Denidni TA (2015) Planar antennas for compact multiband transceivers using a microstrip feedline and multiple open-ended ground slots. IET Microw Antennas Propag 9(5):486–494
11. Das S, Chowdhury P, Biswas A, Sarkar PP, Chowdhury SK (2015) Analysis of a miniaturized multiresonant wideband slotted microstrip antenna with modified ground plane. IEEE Antennas Wirel Propag Lett 14:60–63

12. Liu Y, Zhou YM, Liu GF, Gong SX (2016) Heptaband inverted-F antenna for metal-rimmed mobile phone applications. IEEE Antennas Wirel Propag Lett 15:996–999
13. Chaabane G, Madrangeas V, Chatras M, Arnaud E, Huitema L, Blondy P (2017) High-linearity 3-bit frequency-tunable planar inverted-F antenna for RF applications. IEEE Antennas Wirel Propag Lett 16:983–986
14. Deng J, Li J, Zhao L, Guo L (2017) A dual-band inverted-F MIMO antenna with enhanced isolation for WLAN applications. IEEE Antennas Wirel Propag Lett 16:2270–2273
15. Asadallah FA, Costantine J, Tawk Y (2018) A multiband compact reconfigurable PIFA based on nested slots. IEEE Antennas Wirel Propag Lett 17(2):331–334
16. Yan S, Volskiy V, Vandenbosch GAE (2017) Compact dual-band textile PIFA for 433-MHz/2.4-GHz ISM bands. IEEE Antennas Wirel Propag Lett 16:2436–2439
17. Behdad N, Sarabandi K (2005) A compact antenna for ultrawide-band applications. IEEE Trans Antennas Propag 53(7):2185–2192
18. Ghaffar FA, Yang S, Cheema HM, Shamim A (2016) A 24 GHz CMOS oscillator transmitter with an inkjet printed on-chip antenna. In: IEEE MTT-S international microwave symposium (IMS), San Francisco, CA, pp 1–3
19. Tang MC, Shi T, Ziolkowski RW (2015) Flexible efficient quasi-Yagi printed uniplanar antenna. IEEE Trans Antennas Propag 63(12):5343–5350
20. Booker HG (1946) Slot aerials and their relation to complementary wire aerials (Babinet's principle). J Inst Electr Eng Part IIIA: Radiolocation 93(4):620–626
21. Dyson J (1959) The unidirectional equiangular spiral antenna. IRE Trans Antennas Propag 7(4):329–334
22. Wei X, Liu J, Long Y (2018) Printed log-periodic monopole array antenna with a simple feeding structure. IEEE Antennas Wirel Propag Lett 17(1):58–61
23. El Hamdouni A et al (2015) A novel design of a CPW-fed printed fractal antenna for UWB applications. In: Third international workshop on RFID and adaptive wireless sensor networks (RAWSN), Agadir, Morocco, pp 38–41
24. Sabban A (2017) New fractal compact printed antennas. In: IEEE international symposium on antennas and propagation and USNC/URSI national radio science meeting, San Diego, CA, pp 2197–2198
25. Gianvittorio JP, Romeu J, Blanch S, Rahmat-Samii Y (2003) Self-similar prefractal frequency selective surfaces for multiband and dual-polarized applications. IEEE Trans Antennas Propag 51(11):3088–3096
26. Visser HJ, Reniers ACF, Theeuwes JAC (2008) Ambient RF energy scavenging: GSM and WLAN power density measurements. In: IEEE 38th European microwave conference, Amsterdam, Netherlands, pp 721–724
27. Arrawatia M, Baghini MS, Kumar G (2010) RF energy harvesting system at 2.67 and 5.8 GHz. In: IEEE Asia-Pacific microwave conference, Yokohama, Japan, pp 900–903
28. Boaventura A, Collado A, Carbalho NB, Georgiadis A (2013) Optimum behavior, wireless power transmission system design through behavioral models and efficient synthesis techniques. IEEE Microw Mag 14(2):26–35
29. Peter TT, Rahman TA, Cheung SW, Nilavalan R, Abutarboush HF, Vilches A (2014) A novel transparent UWB antenna for photovoltaic solar panel integration and RF energy harvesting. IEEE Trans Antennas Propag 62(4):1844–1853
30. Arrawatia M, Baghini MS, Kumar G (2011) RF energy harvesting from cell towers in 900 MHz band. In: IEEE national conference on communications (NCC), Bangalore, India, pp 1–5
31. EnOcean website (2014) Energy harvesting wireless technology. Available at: https://www.enocean.com/en/energy-harvesting-wireless/
32. NikkoIA website (2014) Applications. Available at: https://www.nikkoia.com/en/bizapplications/security/
33. Voltree Power website (2014) Bioenergy harvester. Available at: https://voltreepower.com/bioHarvester.html
34. Georgiadis A, Andia G, Collado A (2010) Rectenna design and optimization using reciprocity theory and harmonic balance analysis for electromagnetic (EM) energy harvesting. IEEE Antennas Wirel Propag Lett 9:444–446

35. European Commission—Digital Agenda for Europe (2014) Innovations in energy harvesting and storage. Available at: https://ec.europa.eu/digital-agenda/futurium/en/content/innovations-energy-harvesting-and-storage
36. Technology Strategy Board, Energy Harvesting: Watts Needed? Workshop, Royal Society, Carlton House Terrace London (2011)
37. Costanzo A, Donzelli F, Masotti D, Rizzoli V (2010) Rigorous design of RF multiresonator power harvesters. In: Fourth European conference on antennas and propagation, Barcelona, Spain, pp 1–4
38. Rahman A, Alomainy A, Hao Y (2007) Compact body-worn coplanar waveguide fed antenna for UWB body-centric wireless communications. In: Second European conference on antennas and propagation, Edinburgh, UK, pp 1–4
39. Hansen RC, Collin RE (2011) Small antenna handbook. Wiley, IEEE Press
40. Yu A, Zhang X (2002) A broadband patch antenna array for wireless LANs. In: IEEE antennas and propagation society international symposium, San Antonio, USA, pp 228–231
41. Ali M, Yang G, Dougal R (2005) A new circularly polarized rectenna for wireless power transmission and data communication. IEEE Antennas Wirel Propag Lett 4:205–208
42. Chung K, Nam Y, Yun T, Choi J (2006) Reconfigurable microstrip patch antenna with switchable polarization. ETRI J 28(3):379–382
43. Borja P, Romeu N (2000) An iterative model for fractal antennas: application to the sierpinski gasket antenna. IEEE Trans Antennas Propag 48(5):713–719
44. Hoang TQV, Douyere A, Dubard JL, Lan Sun Lunk JD (2011) TLM design of a compact PIFA rectenna. In: International conference on electromagnetics in advanced applications (ICEAA), Torino, Italy, pp 508–511
45. Hyok JS, Tsung YH, Sievenpiper DF, Hui PH, Schaffner J, Yasan E (2008) A method for improving the efficiency of transparent film antennas. IEEE Antennas Wirel Propag Lett 7:753–756
46. Paing T, Shin J, Zane R, Popovic Z (2008) Resistor emulation approach to low-power RF energy harvesting. IEEE Trans Power Electron 23(3):1494–1501
47. Masotti D, Constanzo A, Adam S (2011) Design and realization of a wearable multifrequency RF energy harvesting system. In: 5th European conference on antennas and propagation (EUCAP), Rome, pp 517–520
48. Weddell AS, Merrett GV, Kazmierski TJ, El-Hashimi BM (2012) Accurate supercapacitor modeling for energy harvesting Wireless Nodes. IEEE Trans Circ Syst II Exp Briefs 58(12):911–915

Chapter 4
Matching Network and Rectifier Circuit

4.1 Introduction

The rectifier and matching circuit play an important role in converting the microwave signal into DC power. To maximize conversion efficiency, the antenna impedance should be perfectly matched with the rectifier impedance, which can be achieved by properly designing the matching circuit. The reflection coefficient is defined by the following relation

$$\Gamma = \frac{Z_{\text{rect}} - Z^*_{\text{ant}}}{Z_{\text{rect}} + Z^*_{\text{ant}}} \tag{4.1}$$

where $Z^*_{\text{ant}} = R_{\text{ant}} + X_{\text{ant}}$ is the conjugate impedance of the antenna and $Z_{\text{rect}} = R_{\text{rect}} + X_{\text{rect}}$ is the input impedance of the rectifier. When the real (R_{ant} and R_{rect}) and reactive components (X_{ant} and X_{rect}) are equal, a perfect matching is obtained.

A rectifier converts AC power into DC power. Various topologies of rectifiers were reported to perform AC-DC rectification. The rectifier's output is not a pure DC voltage, so a DC pass filter is used in combination with the rectifier circuit to remove or suppress the ripples in the output voltage. This chapter describes the various structures used to implement the rectifier and matching circuit.

4.2 Distributed and Lumped Circuits

Distributed circuits are composed of distributed components and transmission lines. As compared to the traditional circuits, which were made of passive components such as inductors, capacitors, resistors, and transformers, the distributed circuits are used at microwave frequencies where lumped components are hard to realize. Distributed circuits can be easily designed on the printed circuit board using cascaded lines,

B. K. Kanaujia et al., *Rectenna: Wireless Energy Harvesting System*,
Advances in Sustainability Science and Technology,
https://doi.org/10.1007/978-981-16-2536-7_4

coupled lines, and stubs. These circuits are widely utilized to design circulators, directional couplers, filters, and power dividers in the microwave range.

4.2.1 Construction

A large proportion of the distributed circuit is made up of transmission lines. The cross-sectional components of the line do not change along its length. Additional simplification occurs in equivalent transmission line circuits, in which all components are of the same length. A lumped circuit model made up of inductors, capacitors, and resistors can be easily converted into a distributed circuit [1].

4.2.2 Advantages and Disadvantages

Distributed-component electrical networks are simple to fabricate, but they occupy more space than lumped-component circuits. They are not preferred for handheld devices, where space is restricted due to the small size of the system. If the working frequency is not very high, the designer may choose to scale down parts rather than switch to distributed components. However, in lumped electrical circuits, the parasitic and resistive components become more prominent. Therefore, designers may choose a distributed-component configuration, regardless of whether lumped parts are available at the time, to benefit from improved quality [2, 3].

4.2.3 Types of Distributed Circuit

(i) Filter
Transmission lines are commonly used in the construction of filters. They are built using cascaded lines, coupled lines, and stubs. Interdigital, comb line, and hairpin filters have been widely used for many years. Latest developments include the use of fractal filters [4].

In lumped-component channels, extra components are utilized. However, the structure becomes extremely complex as the filter approaches an ideal response [5]. A single resonator (e.g., a stub or spur line) may be enough for simple, narrow band requirements [6]. Figure 4.1 shows a schematic of a microwave low-pass stub filter.

(ii) Power divider, combiner, and directional coupler
A directional coupler, shown in Fig. 4.2, is a multi-port microwave device that transfers power from one port to another. The power that enters the input port is coupled to a third port known as the coupled port. None of the power is

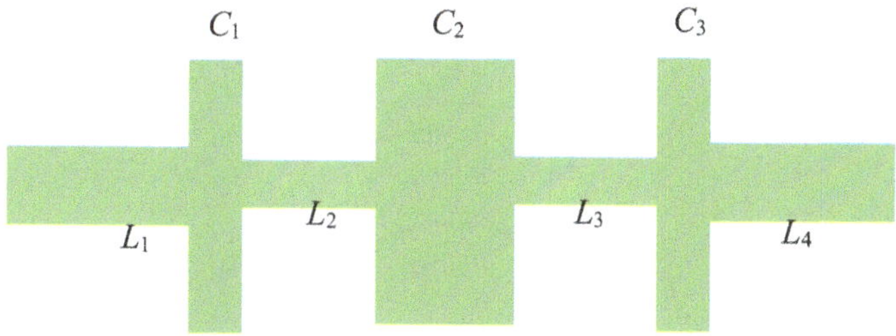

Fig. 4.1 Microwave low-pass stub filter

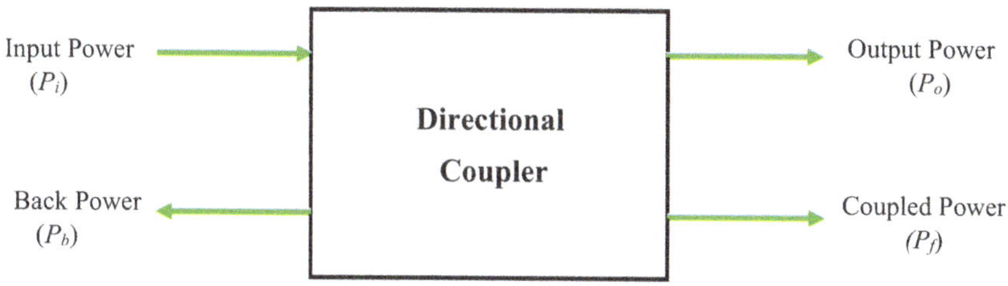

Fig. 4.2 Directional coupler [10]

coupled to the fourth port, which is commonly referred to as the isolated port [7].

A power divider can also function as a directional coupler, with the detached port terminating as a matched load (making it effectively a three-port device). When the coupling factor (the degree of force visible at the coupled port) is high, the title "power divider" is commonly used, whereas when the coupling factor is low, the title "directional coupler" is used [8, 9].

(iii) Hybrid
A hybrid circuit evenly distributes power between the coupled and output ports [10]. Although the term "hybrid" originally referred to the hybrid transformer (used in phones), it now has a broader meaning. A hybrid ring or rat-race coupler is a hybrid circuit that does not include any coupled lines. Each hybrid port is connected to a transmission line ring. Standing waves are caused by waves traveling in the opposite direction around the rings [11]. Figure 4.3 shows a schematic of a hybrid ring device.

(iv) Circulator
A circulator is a multi-port (three- or four-port) device that transfers input power to the adjacent port. Power can move clockwise or anticlockwise, but it only moves in one direction. The rotation of the circulator is determined by Faraday's rotation [12, 13]. Generally, circulators are used in reflection amplifiers, where the negative resistance of the Gunn diode is used to reflect more power than the received [14].

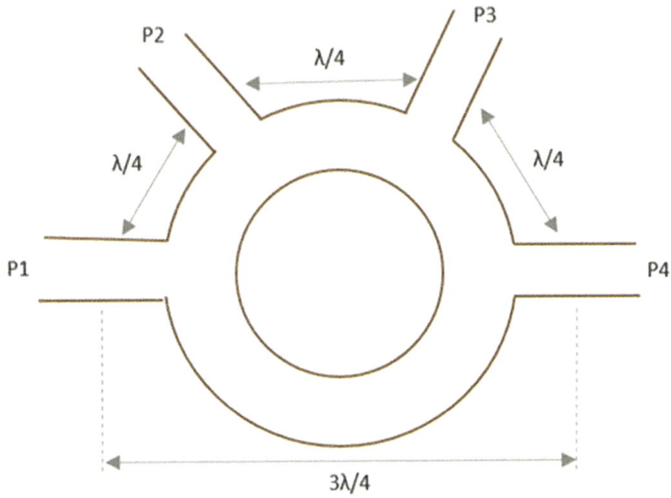

Fig. 4.3 Hybrid ring microwave device

Circulators, unlike other reciprocal lumped and distributed circuits, are a special case [15]. A reciprocal circuit has symmetrical S-parameters. Figure 4.4 shows a schematic of a three-port circulator.

Fig. 4.4 Three-port circulator

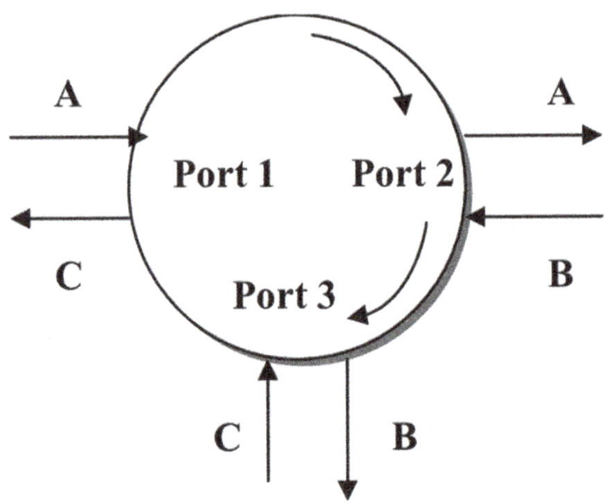

4.3 Matching Network

The impedance matching network matches the source and load impedances. This process is also known as tuning. The impedance matching is needed due to the following reasons.

- A maximum power is transferred when the source and load are perfectly matched.
- In a power distribution network, like an antenna array, the amplitude and phase errors can be reduced by using a matching circuit.
- In a receiver system, comprising of a low-noise amplifier and antenna, the signal-to-noise ratio can be improved by the impedance matching network.

Figure 4.5 gives a basic idea of the matching network, which is placed between the source and the load. The maximum power transfer theorem can be used to transfer the most power from the source to the load. There are various types of matching networks available, which are chosen based on the following criteria as shown in Fig. 4.6.

i. **Bandwidth**: It is simple to match impedance at a single frequency. But, in many applications, impedance matching at multiple frequencies and over a band of frequencies is required. There are various ways to achieve matching over a wide frequency band and at multiple frequencies.
ii. **Complexity**: A less complex impedance matching network is required because it has several advantages over any other complex structure, including lower cost, smaller size, less lossy, and higher reliability.
iii. **Adjustability**: An adjustable matching network can match the impedance for a variety of load values. Such types of matching networks are preferred for the variable loads.
iv. **Implementation**: In some applications, the matching network is determined by the type of circuit that needs to be matched; for example, for transmission line matching, a tuning stub or quarter-wave transformer is preferred over a multi-section transformer or L-network matching circuit.

Figure 4.7 depicts the topology of a matching network.

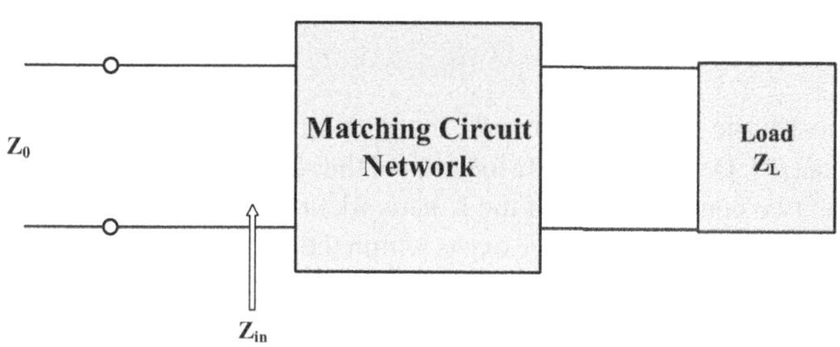

Fig. 4.5 Matching network

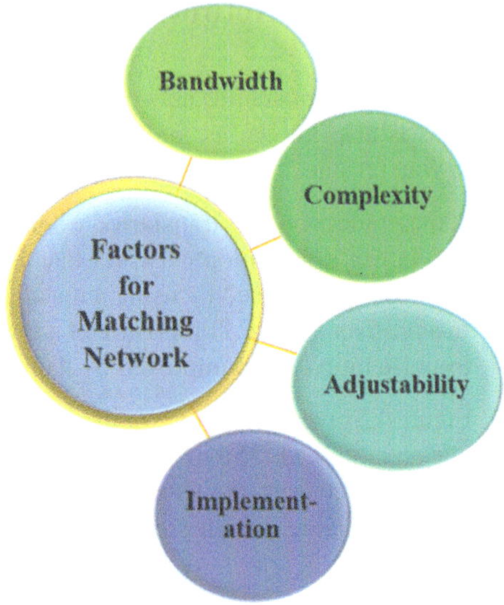

Fig. 4.6 Factors to consider when choosing a matching network circuit

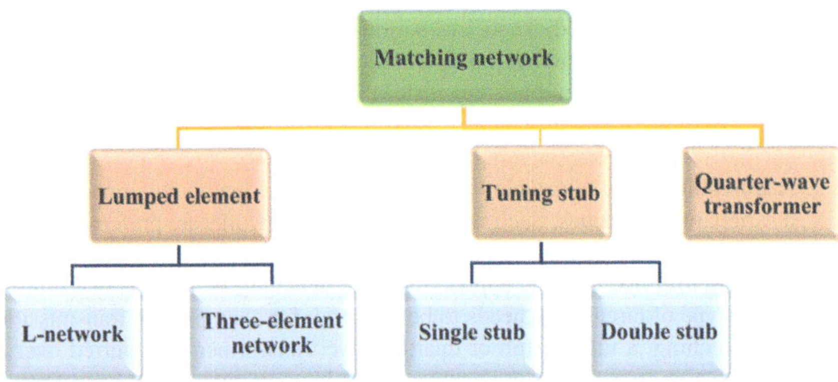

Fig. 4.7 Matching network topology

4.3.1 L-Network

It is the most basic matching network, made up of series reactance (jX) and shunt susceptance (jB). Depending on the location of the shunt elements, shown in Figs. 4.8 and 4.9, the two configurations of the L-network are widely used.

The normalized load impedance exists within the circle $(1 + jX)$ on the smith chart in the L-network, as shown in Fig. 4.8, resulting in $R_L > Z_0$. The load impedance is a complex conjugate

$$Z_L = R_L + jX_L \tag{4.2}$$

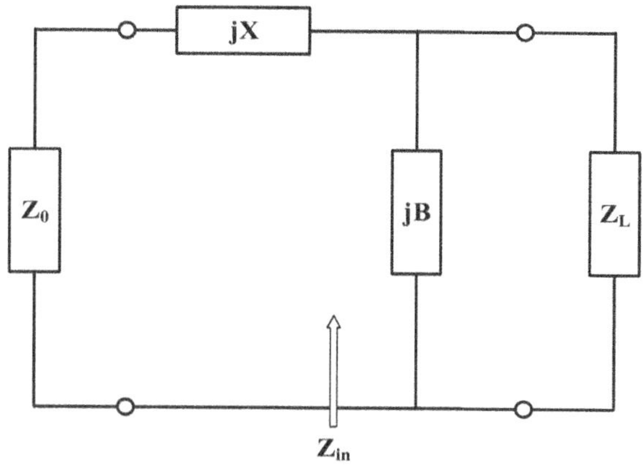

Fig. 4.8 L-network ($R_L > Z_0$)

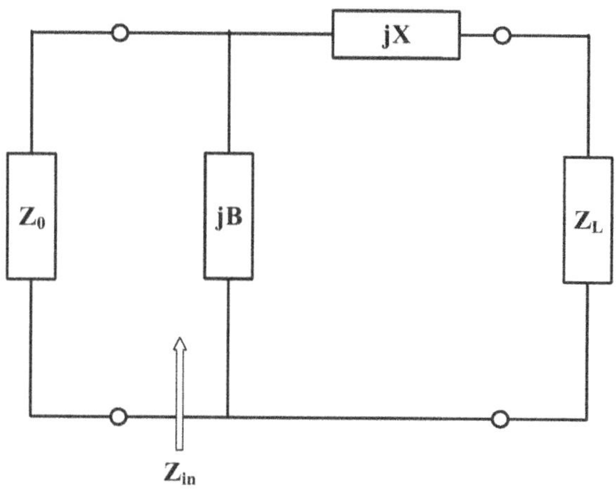

Fig. 4.9 L-network ($R_L < Z_0$)

The input impedance followed by the load impedance matching condition is

$$Z_{\text{in}} = jX + \frac{1}{jB + \frac{1}{R_L + jX_L}} \tag{4.3}$$

$$Z_0 = Z_{\text{in}} \tag{4.4}$$

The B and X can be solved by rearranging and separating the imaginary and real parts

$$B(XR_L - X_L Z_0) = R_L - Z_0 \tag{4.5}$$

$$X(1 - BX_L) = BZ_0 R_L - X_L \tag{4.6}$$

On solving the above two equations, Eq. (4.5) gives the expression for B

$$B = \frac{X_L \pm \sqrt{R_L/Z_0} \cdot \sqrt{R_L^2 + X_L^2 + Z_0 R_L}}{R_L^2 + X_L^2} \tag{4.7}$$

The X can be determined by using Eq. (4.6)

$$X = \frac{B Z_0 R_L - X_L}{1 - B X_L} \tag{4.8}$$

Equation (4.7) indicates that there can be two solutions for X and B and both are physically realizable.

In Fig. 4.9, the L-network is designed for $R_L < Z_0$, which implies that the normalized load impedance exists outside the circle $(1 + jX)$ of the smith chart.

For perfect matching, the load impedance must be equal to the $\frac{1}{Z_0}$

$$Z_L = R_L + jX_L \tag{4.9}$$

The admittance for impedance matched condition is

$$\frac{1}{Z_0} = jB + \frac{1}{R_L + j(X + X_L)} \tag{4.10}$$

$$Z_0 = Z_{in} \tag{4.11}$$

After solving the above equation and separating the real and imaginary parts

$$B Z_0 (X + X_L) = Z_0 - R_L \tag{4.12}$$

$$(X + X_L) = B Z_0 R_L \tag{4.13}$$

X and B can be solved by using the Eqs. (4.12) and (4.13).

$$B = \pm \frac{\sqrt{(Z_0 - R_L)/R_L}}{Z_0} \tag{4.14}$$

$$X = \pm \sqrt{R_L((Z_0 - R_L)} - X_L \tag{4.15}$$

To match the source and complex load impedances in an L-network, the Z_0 must be real and its imaginary part must be zero. The L-type matching network performs well at low frequencies of up to 1 GHz.

4.3.2 *Three-Element Matching Network*

In a three-element matching network, the parameter Q can be selected to add more flexibility to the circuit, which is a limitation in L-network. The three-element matching network has two configurations: Π-matching network and T-matching network.

i. **Π-Matching Network**
   It contains three elements arranged in a Π form as shown in Fig. 4.10. A Π-section is formed by connecting the two configurations of the L-network (in series) as illustrated in Fig. 4.11.

The Z_{in} can be chosen arbitrarily with the help of Q. The first section of the Π circuit matches the Z_{in} impedance, and the second section matches the load impedance. Z_{in} must be less than the source R_S and load R_L because they are connected in a series arm and must satisfy the L-match conditions.

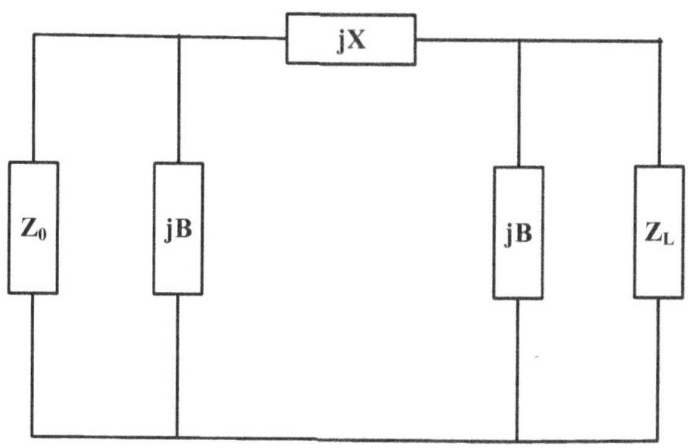

Fig. 4.10 Π-matching network (where $X = X_1 + X_2$)

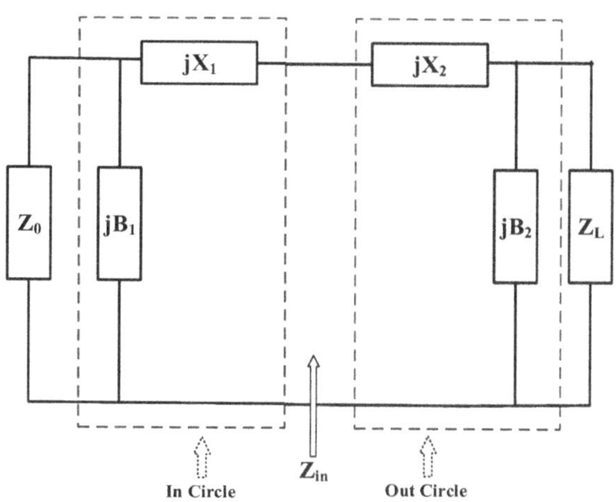

Fig. 4.11 Π-matching network (formed by cascading two L-matching networks)

ii. **T-Match Network**

The T-section can be formed by cascading the two L-type networks as shown in Fig. 4.12. This configuration is dual to the Π-matching network. Out-circle matching is in series with the in-circle matching network; therefore, the susceptibility becomes $B = B_1 + B_2$. The impedance Z_{in} can be chosen in the same manner as the Π-section circuit. In this case, Z_{in} should be greater than the load and source impedance. Here, Q is a combination of the two stages; hence, the value of Q is higher than the single L-section network as shown in Fig. 4.13.

The Q of the three-element circuit is high, which means the bandwidth of the circuit is small. Hence, the T-configuration is used for narrowband applications. As compared to the L-match network, its bandwidth is small due to the fixed value of the Q.

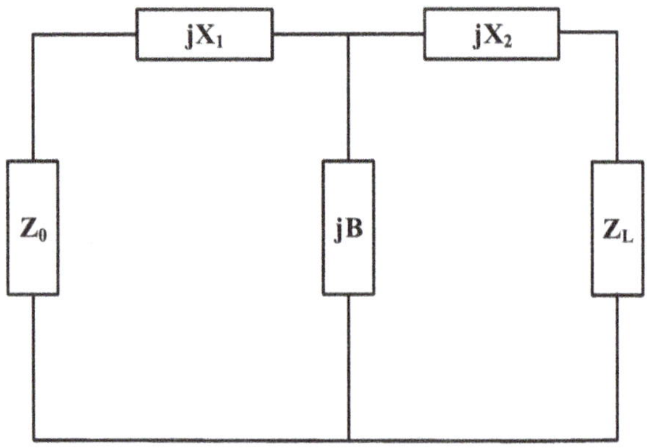

Fig. 4.12 T-matching network

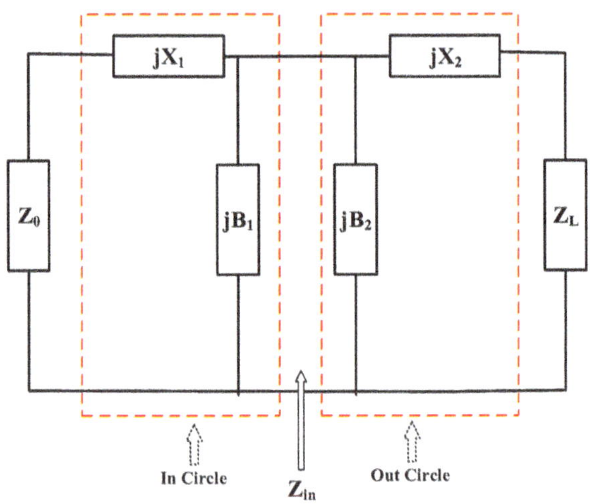

Fig. 4.13 T-matching network (formed by cascading two L-matching networks)

4.3.3 Tuning Stub

This method is widely used for perfect matching. It can easily be fabricated as a part of the transmission line and can avoid the use of lumped elements. Tuning stubs are the part of short-ended open-ended transmission lines and can be used in both ways, short and open stubs for perfect matching. These configurations are generally called distributed impedance matching circuits.

i. **Single Stub Tuning**
In this technique, a T-segment is connected between the source and the load. The characteristic impedance of the transmission line, of length *L*, is the same as the impedance of the source. A short or open stub is connected, at a distance *d* from the load, to cancel the imaginary part and match the resistive part. The open and short stubs can be connected in: (i) parallel or shunt stub, and (ii) series stub as shown in Fig. 4.14.

(a) Shunt stub: The distance *d* from the load is selected so that Y equals $Y + jB$. Then, $-jB$ is selected as the stub susceptance that will cancel the imaginary part.

The *d* and *l* parameters are important for impedance matching, and to derive their formulas the load impedance is assumed to be $Z_L = R_L + jX_L$.

The impedance between the load and the line is

$$Z = Z_0 \frac{(R_L + jX_L) + Z_0 t}{Z_0 + j(R_L + jX_L)t} \tag{4.16}$$

The admittance is

$$Y = G + jB = \frac{1}{Z} \tag{4.17}$$

where $t = \tan \beta d$

$$G = \frac{R_L(1 + t^2)}{R_L^2 + (X_L + Z_0 t)^2} \tag{4.18}$$

$$B = \frac{R_L^2 t - (Z_0 - X_L t)(X_L + Z_0 t)}{Z_0\left[R_L^2 + (X_L + Z_0 t)^2\right]} \tag{4.19}$$

To solve the quadratic equation for *t*, choose a value for *d* such that *G* equals

$$G = Y_0 = \frac{1}{Z_0} \tag{4.20}$$

$$Z_0(R_L - Z_0)t^2 - 2X_L Z_0 t + \left(R_L Z_0 - R_L^2 - X_L^2\right) = 0 \tag{4.21}$$

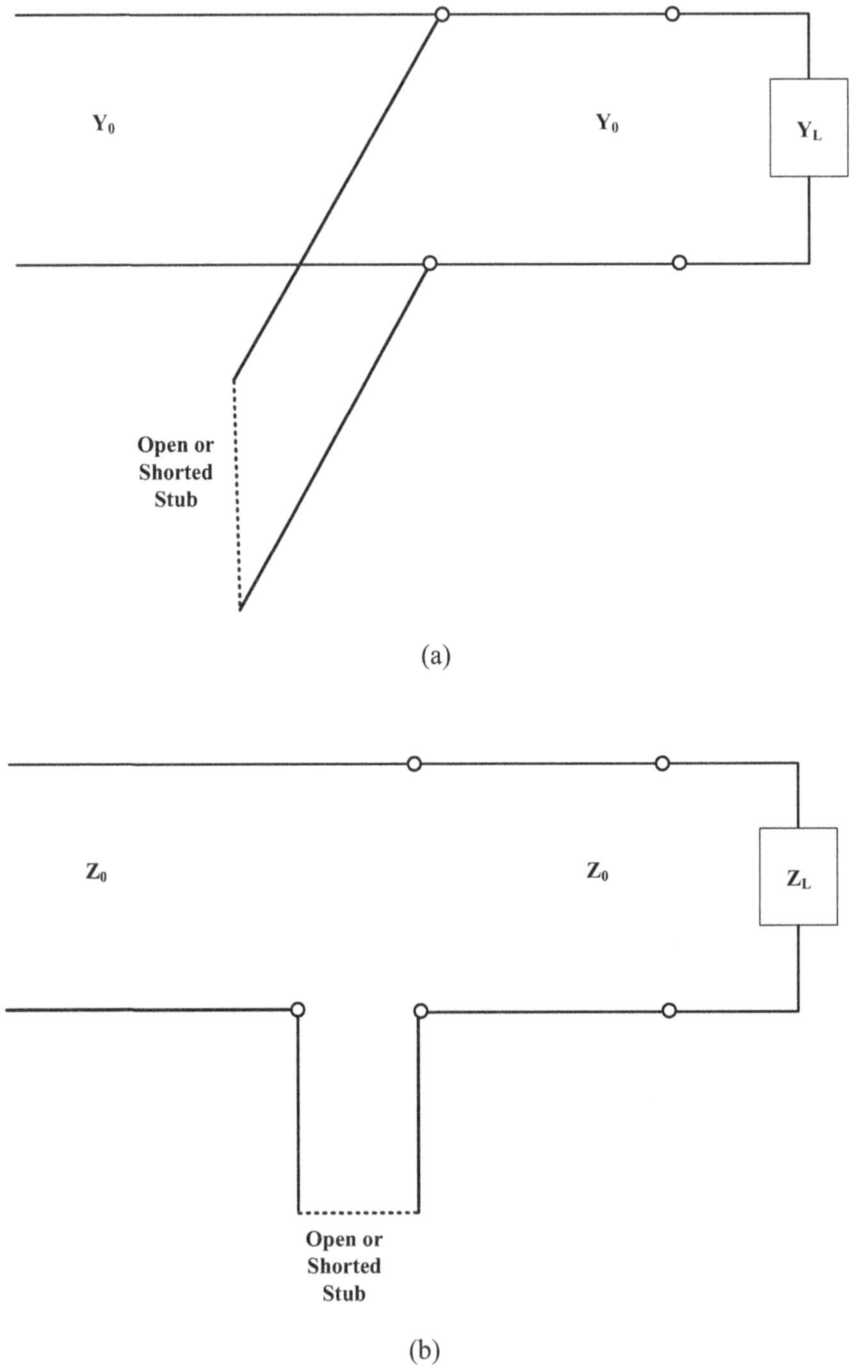

Fig. 4.14 Single stub matching network **a** shunt stub, **b** series stub

The t can be evaluated as

$$t = \frac{X_L \pm \sqrt{R_L\left[(Z_0 - R_L)^2 + X_L^2\right]/Z_0}}{R_L - Z_0} \text{ for } R_L \neq Z_0 \tag{4.22}$$

The two solutions of d can be possible if $t = -X_L/2Z_0$ and $R_L = Z_0$

$$\frac{d}{\lambda} = \begin{cases} \frac{1}{2\pi}\tan^{-1} t & \text{for } t \geq 0 \\ \frac{1}{2\pi}\left(\pi + \tan^{-1} t\right) & \text{for } t < 0 \end{cases} \tag{4.23}$$

$B_S = -B$ is used to find the stub susceptance.
For a short-circuited stub, the length of the stub is

$$\frac{l_S}{\lambda} = \frac{-1}{2\pi}\tan^{-1}\left(\frac{Y_0}{B_S}\right) = \frac{1}{2\pi}\tan^{-1}\left(\frac{Y_0}{B}\right) \tag{4.24}$$

For an open-circuited stub, the stub length can be determined as

$$\frac{l_0}{\lambda} = \frac{1}{2\pi}\tan^{-1}\left(\frac{B_S}{Y_0}\right) = \frac{-1}{2\pi}\tan^{-1}\left(\frac{B}{Y_0}\right) \tag{4.25}$$

(b) Series stub: Similar to the case of a shunt, the distance d from the load is chosen to adjust the Z in the form of $Z + jX$. For perfect matching, the imaginary part stub reactance value should be $-jX$.

The d and l are two important parameters for impedance matching, and to derive their formulas, the load admittance is assumed as $Y_L = G_L + jB_L$.
The admittance Y from the load to the line is

$$Y = Y_0\frac{(G_L + jB_L) + jY_0t}{Y_0 + jt(G_L + jB_L)} \tag{4.26}$$

where $t = \tan\beta$d.

$$Z = R + jX = \frac{1}{Y} \tag{4.27}$$

$$R = \frac{G_L\left(1 + t^2\right)}{G_L^2 + (B_L + Y_0t)^2} \tag{4.28}$$

$$X = \frac{G_L^2t - (Y_0 - B_Lt)(B_L + tY_0)}{Y_0\left[G_L^2 + (B + Y_0t)^2\right]} \tag{4.29}$$

To solve the quadratic equation for t, choose the value of d so that R equals

$$R = Z_0 = \frac{1}{Y_0} \tag{4.30}$$

$$Y_0(G_L - Y_0)t^2 - 2B_L Y_0 t + \left(G_L Y_0 - G_L^2 - B_L^2\right) = 0 \tag{4.31}$$

The t can be evaluated as

$$t = \frac{B_L \pm \sqrt{G_L\left[(Y_0 - G_L)^2 + B_L^2\right]/Y_0}}{G_L - Y_0} \text{ For, } G_L \neq Y_0 \tag{4.32}$$

The two solutions of d can be possible if $t = -B_L/2Y_0$ and $G_L = Y_0$

$$\frac{d}{\lambda} = \begin{cases} \frac{1}{2\pi}\tan^{-1} t & \text{for } t \geq 0 \\ \frac{1}{2\pi}\left(\pi + \tan^{-1} t\right) & \text{for } t < 0 \end{cases} \tag{4.33}$$

$X_S = -X$ is used to find the reactance.
For open-circuited stub, the length of the stub is

$$\frac{l_0}{\lambda} = \frac{-1}{2\pi}\tan^{-1}\left(\frac{Z_0}{X_S}\right) = \frac{1}{2\pi}\tan^{-1}\left(\frac{Z_0}{X}\right), \tag{4.34}$$

For short-circuited stub, the stub length can be determined as

$$\frac{l_S}{\lambda} = \frac{1}{2\pi}\tan^{-1}\left(\frac{X_S}{Z_0}\right) = \frac{-1}{2\pi}\tan^{-1}\left(\frac{X}{Z_0}\right) \tag{4.35}$$

A single stub faces the problem of a variable matching load because it has a fixed length between the source and load, but it can be suitable for a fixed matching circuit.

ii. **Double Stub Tuning**
A double stub tuner solves the variable load problem. In this configuration, two tuning stubs are used at pre-determined locations. The distance between the first stub and the load is variable, and it has the same admittance as the real part admittance of the load. As shown in Fig. 4.15, the length of the first stub must be chosen so that the second stub location has the same real part of admittance as the transmission line admittance. The length of the second stub is chosen to cancel out the imaginary part.

The stub admittance is

$$Y_1 = G_L + j(B_L + B_1) \tag{4.36}$$

B_1 is the first stub susceptance and $Y_1 = G_L + j(B_L + B_1)$ is the load admittance.

The admittance of the second stub is

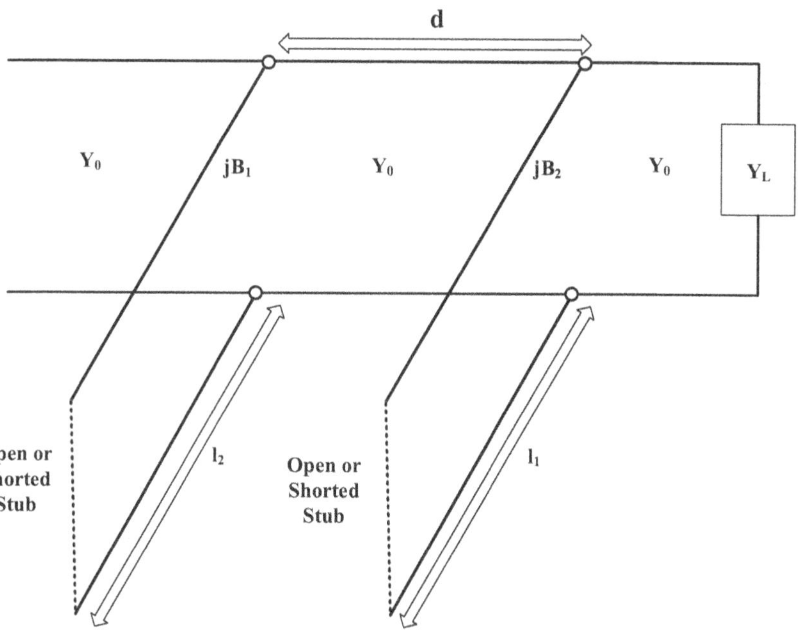

(a)

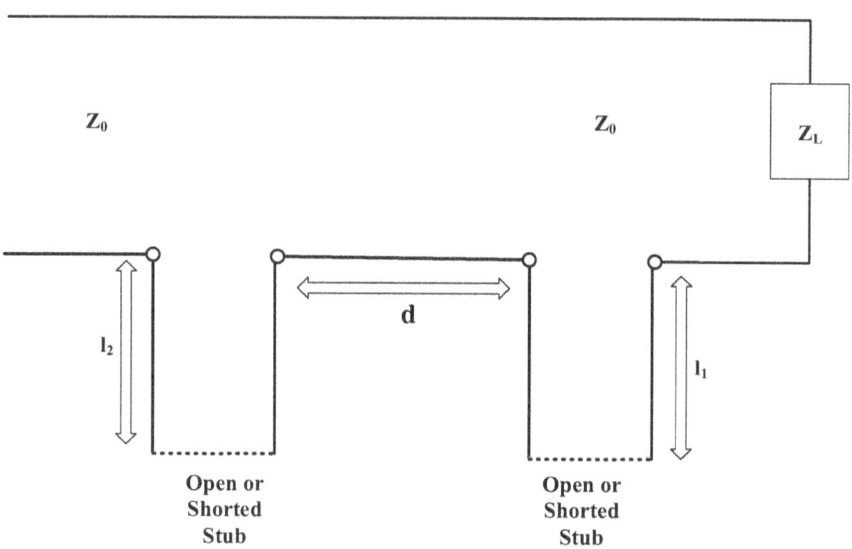

(b)

Fig. 4.15 Double stub matching network **a** shunt stub, **b** series stub

$$Y_2 = Y_0 \frac{G_L + j(B_L + B_1 + Y_0 t)}{Y_0 + jt(G_L + jB_L + jB_1)} \tag{4.37}$$

at $t = \tan \beta d$ and $Y_0 = \frac{1}{Z_0}$.

The real part of Y_2 becomes equal to Y_0 giving the quadratic equation

$$G_L^2 - G_L Y_0 \frac{1+t^2}{t^2} + \frac{(Y_0 - B_L t - B_1 t)^2}{t^2} = 0 \tag{4.38}$$

The G_L can be solved as

$$G_L = Y_0 \frac{1+t^2}{t^2}\left[1 \pm \sqrt{1 - \frac{4t^2(Y_0 - B_L t - B_1 t)^2}{Y_0^2\left(1+t^2\right)^2}}\right] \tag{4.39}$$

$$0 \le \frac{4t^2(Y_0 - B_L t - B_1 t)^2}{Y_0^2\left(1+t^2\right)^2} \le 1 \tag{4.40}$$

Equation (4.40) must be non-negative as G_L is real

$$0 \le G_L \le Y_0 \frac{1+t^2}{t^2} = \frac{Y_0}{\sin^2 \beta d} \tag{4.41}$$

Equation (4.41) provides the range of G_L, and the susceptibility of the first stub can be determined from (4.38)

$$B_1 = -B_L + \frac{Y_0 \pm \sqrt{\left(1+t^2\right) G_L Y_0 - G_L^2 t^2}}{t} \tag{4.42}$$

The negative of the imaginary part of the equation can be used to calculate the susceptibility of the second stub (4.37)

$$B_2 = \frac{\pm Y_0 \pm \sqrt{\left(1+t^2\right) G_L Y_0 - G_L^2 t^2} + G_L Y_0}{G_L t} \tag{4.43}$$

By solving the above two equations, the open-circuited stub length is calculated as

$$\frac{l_S}{\lambda} = \frac{-1}{2\pi} \tan^{-1}\left(\frac{Y_0}{B}\right) \tag{4.44}$$

The short-circuited stub length is

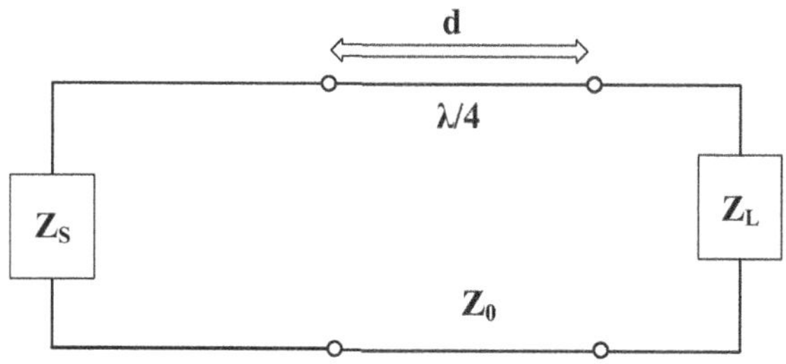

Fig. 4.16 Quarter-wave transformer

$$\frac{l_0}{\lambda} = \frac{1}{2\pi}\tan^{-1}\left(\frac{B}{Y_0}\right) \tag{4.45}$$

where B is equal to B_1 or B_2.

iii. **Quarter-wave Transformer**
A quarter-wave transformer is the simple and most reliable matching network in the distributed circuits. It is an intermediate section of length $d = \lambda/4$ between two systems of different impedances, as shown in Fig. 4.16. The two systems could be either a source or a load. A single-section transformer can only match the frequency at $\lambda/4$, and a multi-section transformer can be used for wideband impedance matching. Furthermore, it can only match the circuit where the load is real. In the case of a complex load, it must be converted into real load impedance using series or shunt reactive components or an appropriate length transmission line.

The impedance of the matching section is

$$Z_0 = \sqrt{Z_s Z_L} \tag{4.46}$$

where Z_S is the source impedance and Z_L is the load impedance. The input impedance of the circuit is

$$Z_{in} = Z_1 \frac{Z_L + jZ_1 t}{Z_1 + jZ_L t} \tag{4.47}$$

When $\beta_l = \theta = \pi/2$ and $t = \tan \beta l$, the reflection coefficient can be calculated as

$$\Gamma = \frac{Z_{in} - Z_0}{Z_{in} + Z_0} = \frac{Z_1(Z_L - Z_0) + jt(Z_1^2 - Z_0 Z_L)}{Z_1(Z_L + Z_0) + jt(Z_1^2 + Z_0 Z_L)} \tag{4.48}$$

$$Z_1^2 = Z_0 Z_L \tag{4.49}$$

Using Eqs. (4.48) and (4.49), the reflection coefficient is

$$\Gamma = \frac{Z_L - Z_0}{Z_L + Z_0 + j2t\sqrt{Z_0 Z_L}} \tag{4.50}$$

The magnitude of the reflection coefficient is

$$\Gamma = \frac{|Z_L - Z_0|}{\left[(Z_L + Z_0)^2 + 4t^2 Z_0 Z_L\right]^{1/2}} \tag{4.51}$$

$$= \frac{1}{\left\{(Z_L + Z_0)^2/(Z_L - Z_0)^2 + \left[4t^2 Z_0 Z_L/(Z_L - Z_0)^2\right]\right\}^{1/2}} \tag{4.52}$$

$$= \frac{1}{\left\{1 + \left[4Z_0 Z_L/(Z_L - Z_0)^2\right] + \left[4Z_0 Z_L t^2/(Z_L - Z_0)^2\right]\right\}^{1/2}} \tag{4.53}$$

$$= \frac{1}{\left\{1 + \left[4Z_0 Z_L/(Z_L - Z_0)^2\right]\sec^2\theta\right\}^{1/2}} \tag{4.54}$$

$$1 + t^2 = 1 + \tan^2\theta = \sec^2\theta \tag{4.55}$$

Because the length of the impedance matching section is $\lambda_0/4$ for a given frequency f_0, this matching technique is based on an approximation method.

4.4 Theory of Rectifier

Rectifiers are the basic building block of the wireless energy harvesting system. A rectifier is a device that converts captured ambient energy from the environment into DC power. The basic rectifier model is shown in Fig. 4.17. In this model, a single diode is connected between the source and the load for rectification. The chosen diode must supply high stability and low ripples.

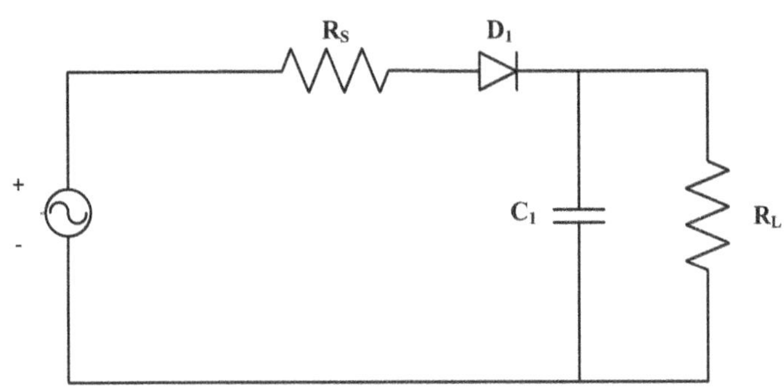

Fig. 4.17 Basic rectifier circuit

4.4.1 Half-Wave Rectifier

A half-wave rectifier is made up of only one diode and can only rectify positive half-cycles. Therefore, it is called a half-wave rectifier or single-way rectifier. A full cycle over the period T is shown in Fig. 4.18. For cycle $0 \rightarrow T/2$, the diode is ON and gives the same output as the input. In this case, the diode acts as a short-circuit equivalent. The diode is turned OFF for the next half-cycle $T/2 \rightarrow t$ and acts as an open-circuit equivalent. The open-circuit provides zero output voltage.

The DC output voltage can be calculated as

$$V_{DC} = \frac{1}{T}\int_0^T V_L(t)\mathrm{d}t \tag{4.56}$$

$$= \frac{1}{2\pi}\int_0^\pi V_s \sin wt\mathrm{d}t \tag{4.57}$$

$$= \frac{V_S}{\pi} = 0.318 \tag{4.58}$$

Other parameters can be calculated as

$$V_L = \sqrt{\frac{1}{T}\int_0^T V_L^2(t)\mathrm{d}t} \tag{4.59}$$

$$= \sqrt{\frac{1}{2\pi}\int_0^\pi Vs^2 \sin^2 wt\mathrm{d}t} = \frac{V_S}{2} \tag{4.60}$$

$$I_{DC} = \frac{V_{DC}}{R_L} = \frac{V_S}{\pi \cdot R_L} \tag{4.61}$$

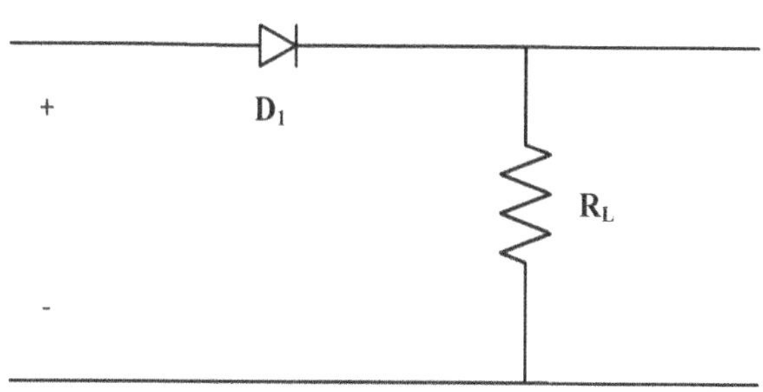

Fig. 4.18 Half-wave rectifier

$$I_L = \frac{V_L}{R_L} = \frac{V_S}{2 \cdot R_L} \tag{4.62}$$

The form factor is defined as

$$\frac{V_L}{V_{DC}} = \frac{\pi}{2} \tag{4.63}$$

The efficiency is calculated as

$$\eta = \left(\frac{1}{\mathrm{FF}}\right)^2 = \frac{4}{\pi^2} = 0.405 \tag{4.64}$$

$$\mathrm{RF} = \sqrt{\mathrm{FF}^2 - 1} = 1.21 \tag{4.65}$$

The peak inverse voltage (PIV) of a diode is an important parameter to consider when designing rectifiers. It is the maximum voltage that a diode can withstand in reverse bias. For half-wave rectification, the PIV is $\geq V_S$.

4.4.2 Full-Wave Rectifier

Center-tapped Transformer

This configuration is made up of two diodes with a tap in the center. Figure 4.19 shows that for positive half-cycle $0 \rightarrow T/2$, diode D_1 is ON and acts as a short-circuit equivalent, while diode D_2 is OFF and acts as an open-circuit equivalent. The rectified voltage is the same for the positive half-cycle. During the next half-cycle, diode D_2 is turned ON and acts as a short-circuit equivalent, while diode D_1 is turned OFF and

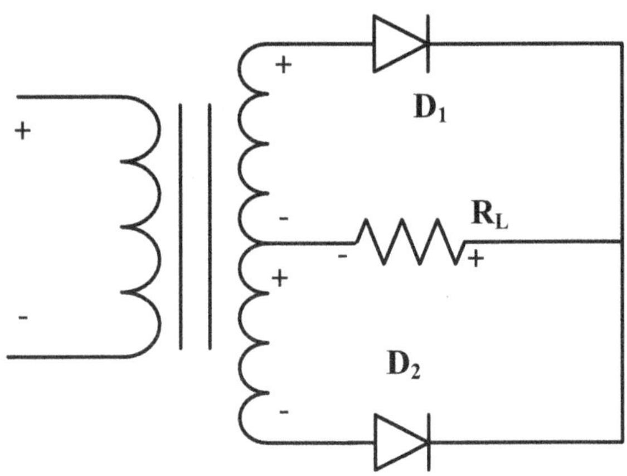

Fig. 4.19 Full-wave rectifier (center-tapped)

acts as an open-circuit equivalent. The cycle and diode operation are reversed for the negative duration, but the same amplitude and polarity as the input are maintained.

The output DC voltage can be determined as

$$V_{\mathrm{DC}} = \frac{1}{T}\int_{0}^{T} VL(t)\mathrm{d}t \tag{4.66}$$

$$= \frac{1}{2\pi}\int_{0}^{\pi} V_{\mathrm{s}} \sin wt\mathrm{d}t = \frac{2VS}{\pi} \tag{4.67}$$

$$V_{\mathrm{L}} = \sqrt{\frac{1}{T}\int_{0}^{T} V_{\mathrm{L}}^{2}(t)\mathrm{d}t} \tag{4.68}$$

$$= \sqrt{\frac{1}{\pi}\int_{0}^{\pi} V_{s}^{2} \sin^{2} wt\mathrm{d}t} = \frac{V_{\mathrm{S}}}{\sqrt{2}} \tag{4.69}$$

$$I_{\mathrm{DC}} = \frac{V_{\mathrm{DC}}}{R_{\mathrm{L}}} = \frac{2V_{\mathrm{S}}}{\pi \cdot R_{\mathrm{L}}} \tag{4.70}$$

$$I_{\mathrm{L}} = \frac{V_{\mathrm{L}}}{R_{\mathrm{L}}} = \frac{V_{\mathrm{S}}}{\sqrt{2} \cdot R_{\mathrm{L}}} \tag{4.71}$$

The form factor is defined as

$$\frac{V_{\mathrm{L}}}{V_{\mathrm{DC}}} = \frac{\pi}{2\sqrt{2}} \tag{4.72}$$

The efficiency is calculated as

$$\eta = \left(\frac{1}{\mathrm{FF}}\right)^{2} = \frac{8}{\pi^{2}} = 0.81 \tag{4.73}$$

$$\mathrm{RF} = \sqrt{\mathrm{FF}^{2} - 1} = 0.483 \tag{4.74}$$

$$\mathrm{PIV} = V_{\mathrm{S}} + V_{\mathrm{S}} \tag{4.75}$$

$$= 2V_{\mathrm{S}} \tag{4.76}$$

A voltage multiplier is a type of electrical network that acts as a rectifier, converting low voltage to high voltage. These circuits are typically made up of capacitors and diodes. Various types of voltage multiplier are available in the open literature. The following section discusses the most commonly used voltage multipliers.

4.4.3 Voltage Doubler

A voltage doubler is an electronic circuit that consists of two capacitors and two diodes connected in series. The capacitor charges through a diode and provides twice the input voltage. Figure 4.20 depicts a single-stage voltage doubler circuit schematic. The diode D_1 is ON during the first half-cycle (positive cycle), charging the capacitor C_1 to the peak output voltage (V_S). The diode D_2 is ON for the next half-cycle (negative cycle), and the diode D_1 is OFF. The diode D_2 conducts and serves as a short-circuit equivalent, charging the capacitor C_2. The charging capacitor charges to the peak value of the output voltage (V_S).

Voltage doublers are available in the following configurations:

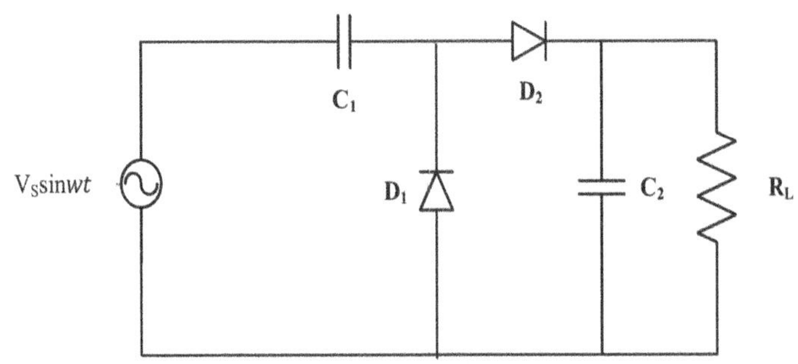

Fig. 4.20 Basic voltage doubler

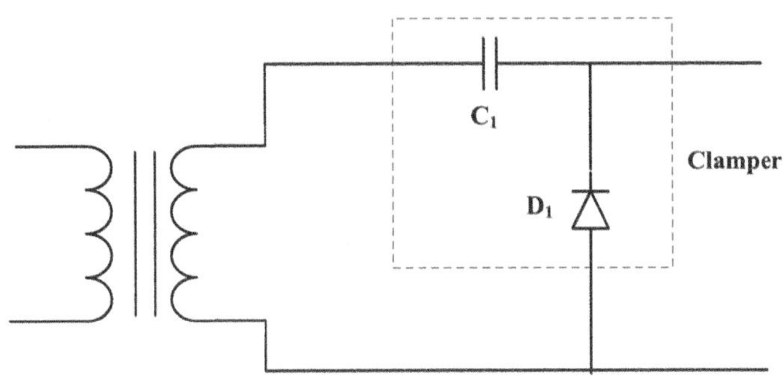

Fig. 4.21 Villard circuit

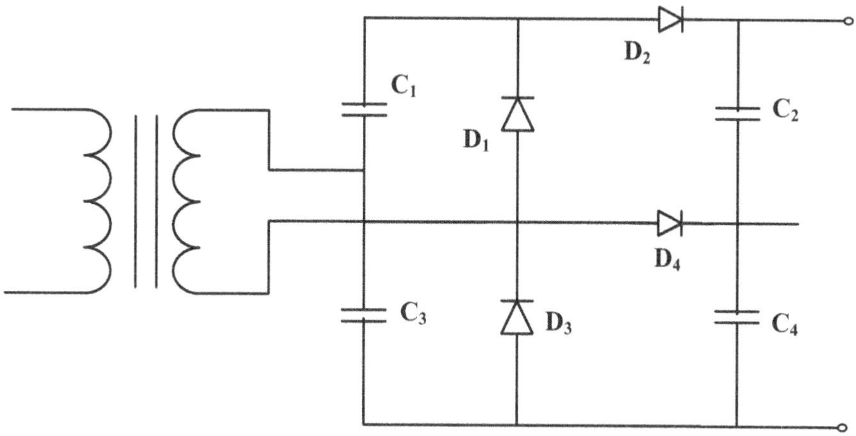

Fig. 4.22 Greinacher rectifier (second order)

i. **Villard Circuit**
Figure 4.21 depicts the simplest voltage doubler, which consists of a diode and a capacitor. The capacitor charges to the peak rectified voltage during the negative half-cycle. This circuit, which acts as a clamper and shifts the rectified DC output, has very poor ripple properties. The AC waves from $-V_S$ to $+V_S$ shifts from 0 V to a peak value of $2V_S$ with the help of a clamper. This double voltage has ripple and cannot be smoothed without the use of an additional filter circuit.

ii. **Greinacher Circuit**
Figure 4.22 depicts another voltage doubler topology known as the Greinacher voltage doubler. It solves the rectified output voltage ripple problem by including some extra components. The posterior part of the circuit acts as an RC filter, removing the ripples. A voltage quadrupler circuit can also be designed by connecting two Greinacher rectifier cells of opposite polarity.

iii. **Delon Circuit**
This rectifier topology is known as a full-wave voltage doubler, and it employs a bridge topology, as illustrated in Fig. 4.23. This circuit was discovered in a cathode ray tube, which was used in television. It provides an extra high tension (EHT) supply to black and white television sets of 10 kV and even more for color television sets. This topology consists of two peak detectors and works in the same way as a basic voltage doubler. The peak detector works for the opposite half-cycle. Since their outputs are connected in series, this results in twice the peak input voltage.

iv. **Voltage Multiplier Circuit**
This topology consists of four diodes and four capacitors as shown in Fig. 4.24. The output DC voltage increases by two, three, four, or many times the input

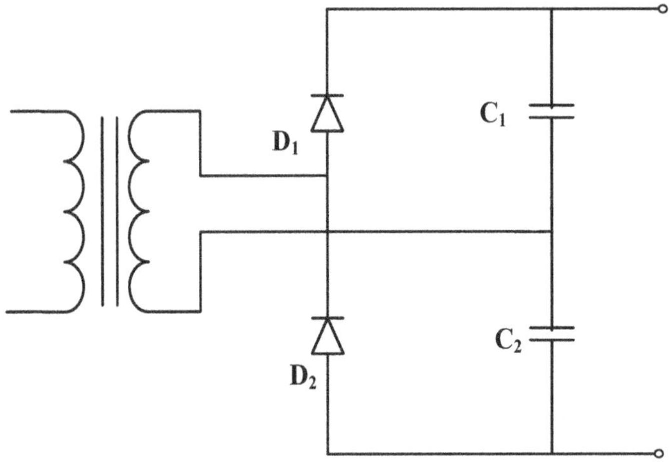

Fig. 4.23 Delon circuit

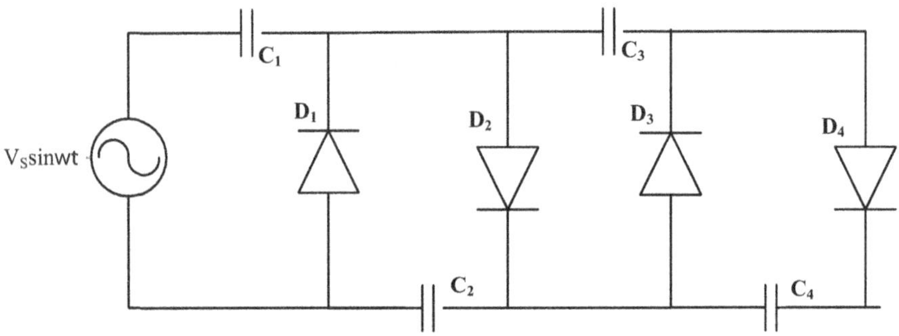

Fig. 4.24 Voltage multiplier circuit

voltage as the number of stages increases. They are referred to as voltage doublers, voltage triplers, voltage quadruplers, and so on. The desired voltage can be achieved by simply cascading the N stages and obtaining the output as $2NV_m$ V. In general, these circuits are used to generate a positive output voltage, but in some applications, a negative voltage is required. Negative voltage can also be obtained by reversing the polarity of the circuit.

4.5 Schottky Diode

The conversion efficiency of the rectenna is determined by the performance of the rectifying diode. There are various types of diodes available for microwave and millimeter ranges, such as point-contact diodes, PIN diodes, tunnel diodes, Schottky diodes, and varactor diodes. Because of the presence of the majority charge carriers, the Schottky diode has many advantages, including low junction capacitance, low junction voltage, and high-speed switching. At microwave frequencies, a Schottky diode behaves almost identically to an ideal diode.

4.5.1 Construction and Working

A Schottky diode is formed by combining the n-type material with the metal electrode. As a result, a metal layer is present on one side of the junction, and an n-type layer is present on the other side.

The Schottky diode symbol and circuit layout are shown in Fig. 4.25. When the circuit is forward biased, current flows from the n-type semiconductor to the metal electrodes due to the movement of electrons. The current generated in the circuit is the unipolar drift current, which is caused by the majority of charge carriers. Since no p-type material is used, the minority charge carriers are not present, leading to a very fast response. The various features of the Schottky diode are shown in Fig. 4.26.

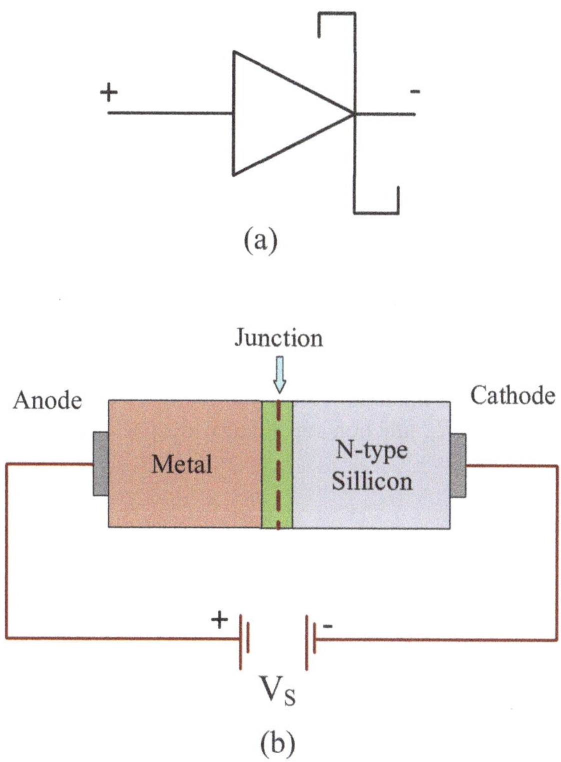

Fig. 4.25 Schottky diode **a** symbol, **b** layout

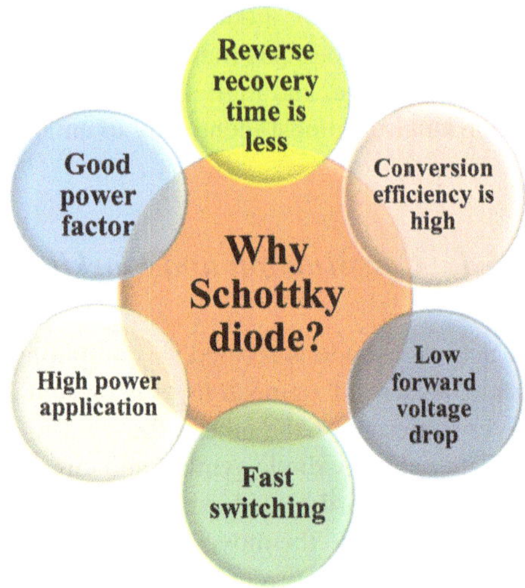

Fig. 4.26 Features of the Schottky diode

The knee voltage or turn-on voltage of the Schottky diode (0.15–0.45 V) is very low as compared to an ordinary p–n junction diode (0.6 V–0.9 V).

4.5.2 Features of the Schottky Diode

Equivalent circuit of the Schottky diode

There are two equivalent circuits for the Schottky diode: high-frequency model and low-frequency model. The high-frequency model, as shown in Fig. 4.27, consists of package capacitance C_P, lead inductance L_L, bond wire inductance L_B, and series resistance junction capacitance C_j.

Figure 4.28 shows the Schottky diode low-frequency (for operation up to 6 GHz) model. It is made up of the following components: series resistance R_S, junction resistance R_j, junction capacitance C_j, package inductor L_P, and package capacitor C_P.

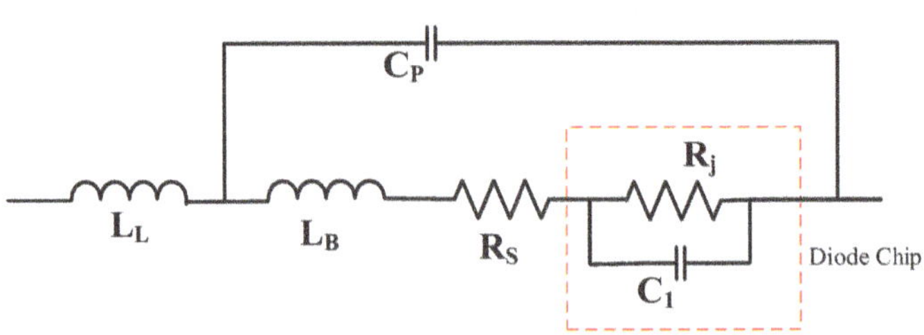

Fig. 4.27 Equivalent circuit of the Schottky diode (high-frequency model)

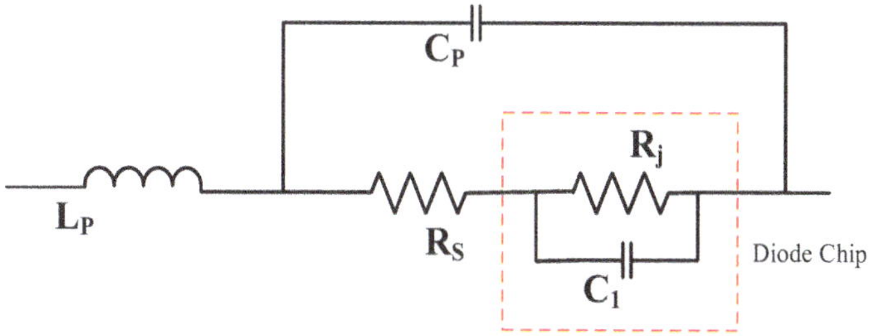

Fig. 4.28 Equivalent circuit of the Schottky diode (low-frequency model)

The following diode parameters must be considered when designing a rectifier circuit.

(i) High reverse leakage current: A high reverse leakage current is advantageous for microwave rectifiers, but it may have an impact on other parameters as well. A Schottky diode has a higher reverse leakage current than a standard p–n junction diode.
(ii) Limited reverse voltage: A Schottky diode has a maximum capacity of 100 V and has a limited reverse voltage due to its structure. These diodes are not suitable for devices that require more reverse voltage.
(iii) Adequate heat sink: A Schottky diode has a low forward voltage drop and does not produce a lot of heat. However, in some applications, such as power rectifiers, heat must be dissipated effectively, necessitating the use of a heat sink.
(iv) Limited junction capacitance: A Schottky diode has a temperature range of 125–175 °C, but one should check the datasheet provided by the manufacturer and use the diodes according to the applications.

References

1. Hunter I (2001) Theory and design of microwave filters. IET, pp 137–130
2. Doumanis E, Goussetis G, Kosmopoulos S (2015) Filter design for satellite communications: helical resonator technology. Artech House, pp 45–46
3. Nguyen C (2015) Radio-frequency integrated-circuit engineering. Wiley, pp 27–28
4. Cohen N (2015) Fractal antenna and fractal resonator primer. In: Michael BM (ed) A life in many dimensions, chap 8. World Scientific, p 220
5. Harrel B (1985) The cable television technical handbook. Artech House, pp 150
6. Awang Z (2013) Microwave systems design. Springer Science and Business Media, pp 296
7. Sisodia ML, Raghuvanshi GS (1987) Basic microwave techniques and laboratory manual. New Age International, p 70
8. Ishii TK (1995) Handbook of microwave technology: components and devices. Academic Press, pp 226

9. Bhat B, Koul SK (1989) "Tripline-like transmission lines for microwave integrated circuits. New Age International, pp 622–627
10. Maloratsky LG (2004) Passive RF and microwave integrated circuits. Elsevier, p 117
11. Chang K, Hsieh L-H (2004) Microwave ring circuits and related structures. Wiley, pp 197–198
12. Sharma KK (2011) Fundamental of microwave and radar engineering. S. Chand Publishing, pp 175–176
13. Linkhart DK (2014) Microwave circulator design. Artech House, p 29
14. Roer TG (2012) Microwave electronic devices. Springer, pp 255–256
15. Maloratsky LG (2004) Passive RF and microwave integrated circuits. Elsevier, pp 285–286

Chapter 5
Rectenna Implementation

5.1 Simulation Using Electromagnetic Simulators

Antennas are used for a wide variety of applications in security, biomedical, wireless communication, wireless power transmission, RFID tags, etc. Verification of design must be performed prior to the fabrication step. The design of the antenna through electromagnetic simulators provides an overview and virtual design to the designer. Also, the desired configuration can be achieved by doing the necessary optimizations in the electromagnetic simulator. Different types of electromagnetic simulation tools are available for designing antenna. This chapter will cover some of the most widely used software/tools.

5.2 High-Frequency Structure Simulator (HFSS)

Ansys HFSS offers a broad range of electromagnetic applications such as designing of antenna even at high frequency with multiple advanced solver technique, designing of feeding system. HFSS simulation is based on different techniques.

(a) Finite element method (FEM).
(b) Physical optics.
(c) FEM transient.
(d) Integral equations.

5.2.1 Finite Element Method (FEM)

A problem can be solved by dividing a geometrical model into smaller parts, and those smaller parts are called finite elements. This system solves the problem of different

B. K. Kanaujia et al., *Rectenna: Wireless Energy Harvesting System*,
Advances in Sustainability Science and Technology,
https://doi.org/10.1007/978-981-16-2536-7_5

fields, including the electromagnetic potential, fluid flow, heat transfer, structural analysis, and mass transport. Some of the advantages of FEM are:

(a) Inclusion of dissimilar material property.
(b) Accurate representation of complex geometry.
(c) Capturing local effects.
(d) Easy representation of the total solution.
(e) Designing a structure.

5.2.2 *Step-by-Step Guide: Antenna Design*

1. Open a new project by clicking on the icon as shown in Fig. 5.1.
2. Select the solution type and choose the driven model (Fig. 5.2).
3. First, design the substrate layer by clicking on the **Draw** toolbar and select the **Box**. Enter the box position and dimensions as shown in the figure below. Select the material from the properties, for example, RT Duroid 5880 with permittivity 2.2, as shown in Fig. 5.3.

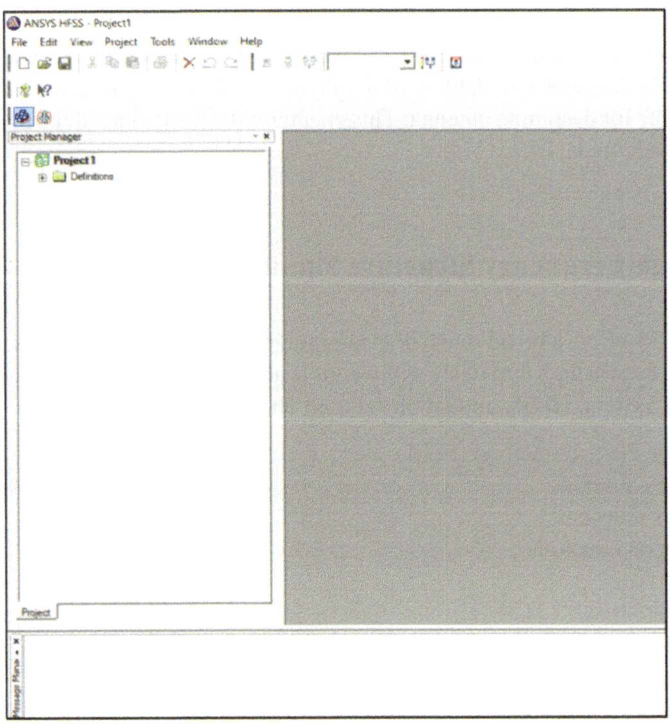

Fig. 5.1 .

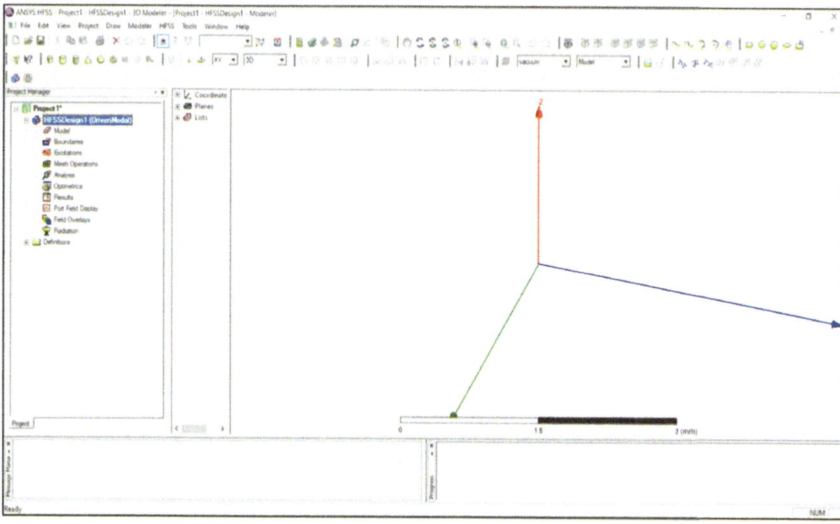

Fig. 5.2 .

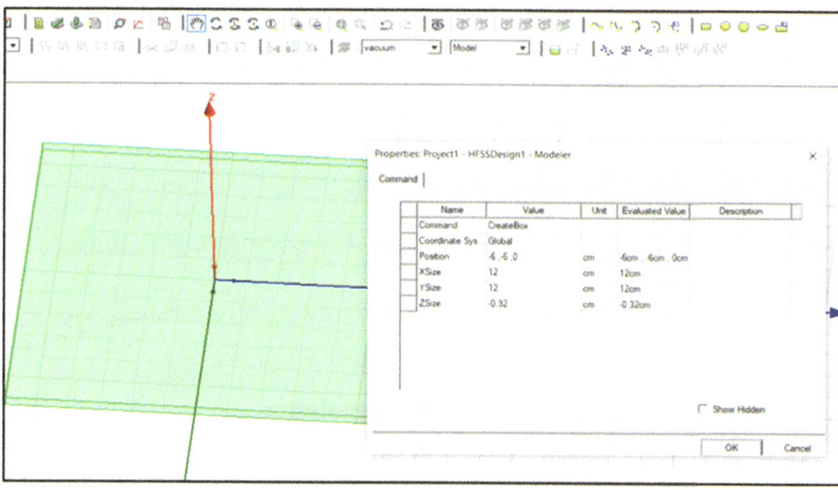

Fig. 5.3 .

4. To define an infinite ground plane, click on the **Draw** -> **Rectangle**, and enter the dimension and location as shown in Fig. 5.4. Then, select the ground plane and assign it as **Perfect E**.
5. To create a circular patch, use **Draw** toolbar > click on **Circle**. Now, add location and radius of the circular patch as shown in Fig. 5.5.
6. A cutout is required for proper excitation at the ground plane. Draw a circle and assign the radius and location accordingly (Fig. 5.6).

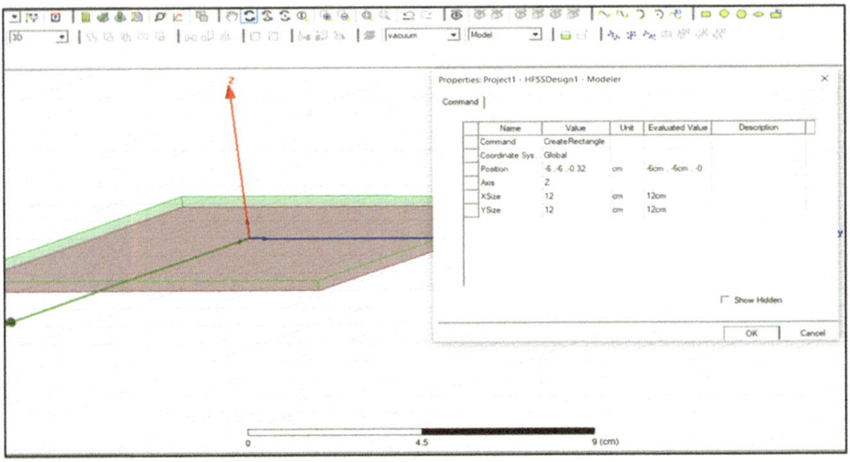

Fig. 5.4 .

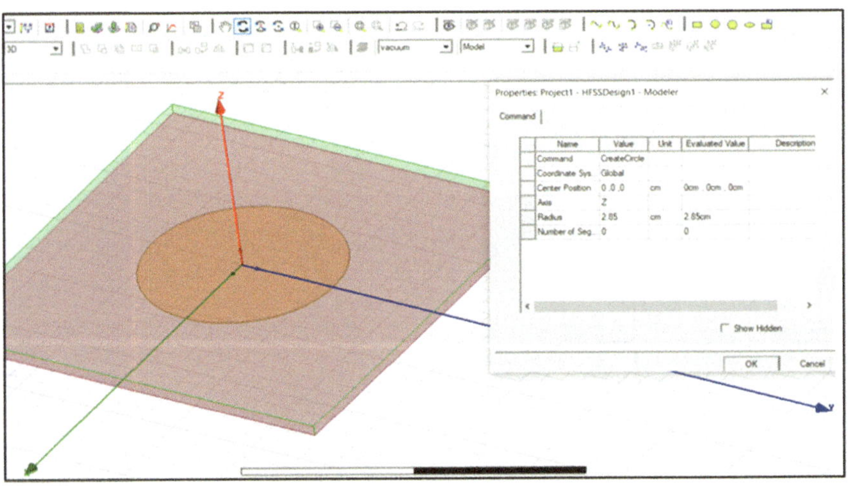

Fig. 5.5 .

7. To create a Coax_Pin, choose a cylinder of radius 7 mm and height 5 mm, as shown in Fig. 5.7. Next, assign it as **PEC** from the material window.
8. As shown in Fig. 5.8, draw a cylinder of radius 7 mm and height 3.2 mm, and name it as probe with material **PEC**.
9. To create an outer cylinder, draw a cylinder of the same dimensions shown in Fig. 5.9. Name it as coax and fill it with a vacuum or Teflon.
10. To assign port excitation, draw a circle at the bottom of the outer cylinder (of the same radius) (Fig. 5.10).

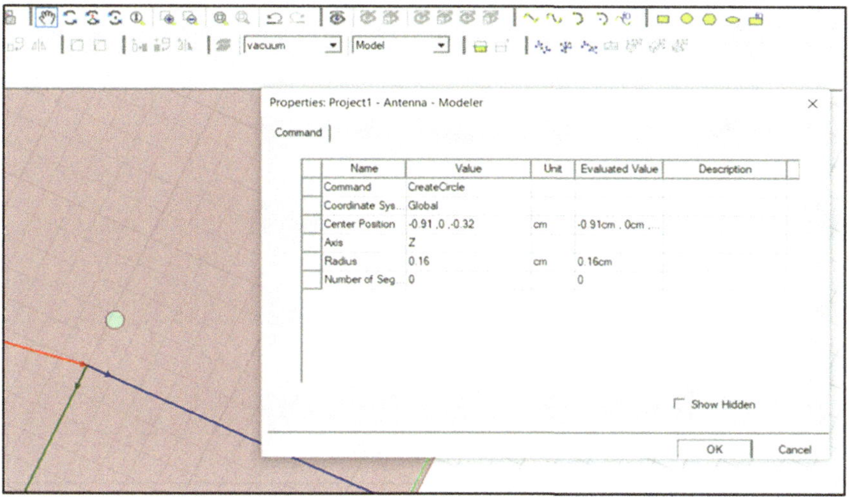

Fig. 5.6 .

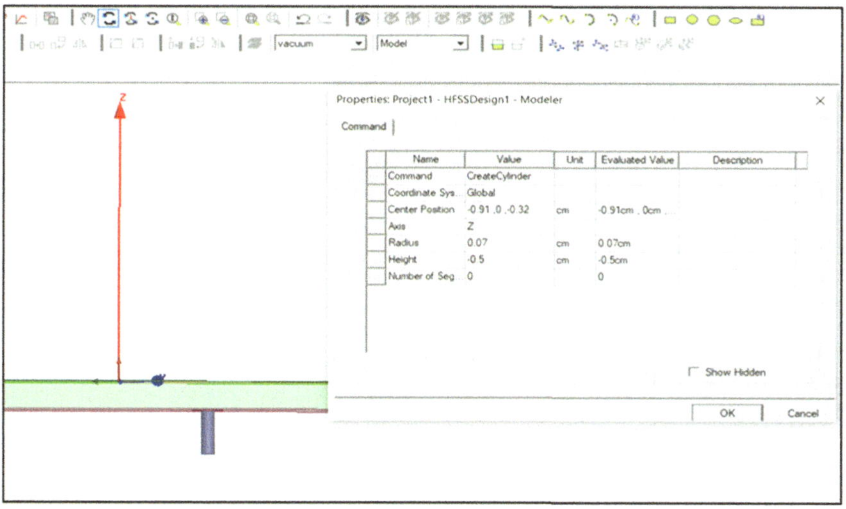

Fig. 5.7 .

11. To assign excitation, click (right) on the circle and select **Waveport** excitation (Fig. 5.11).
12. Add sweep, as mentioned in Fig. 5.12, by clicking on the frequency sweep icon.
13. Validate the design by clicking on the icon shown in Fig. 5.13. This parameter checks the validity of the design.
14. Simulate the project by clicking on the simulation icon (**!**) (Fig. 5.14).

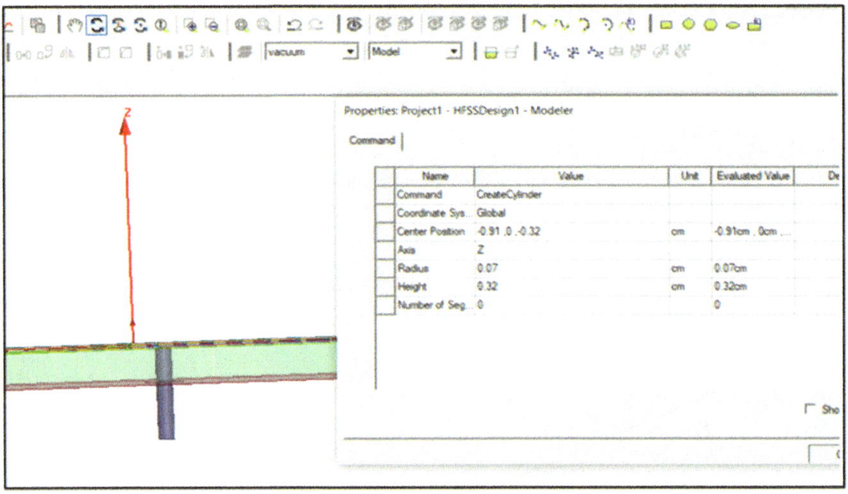
vacuum
Model
Z
Properties: Project1 - HFSSDesign1 - Modeler
Command
Name
Value
Unit
Evaluated Value
Command
CreateCylinder
Coordinate Sys...
Global
Center Position
-0.91 ,0 ,-0.32
cm
-0.91cm , 0cm ,...
Axis
Z
Radius
0.07
cm
0.07cm
Height
0.32
cm
0.32cm
Number of Seg...
0
0

Fig. 5.8 .

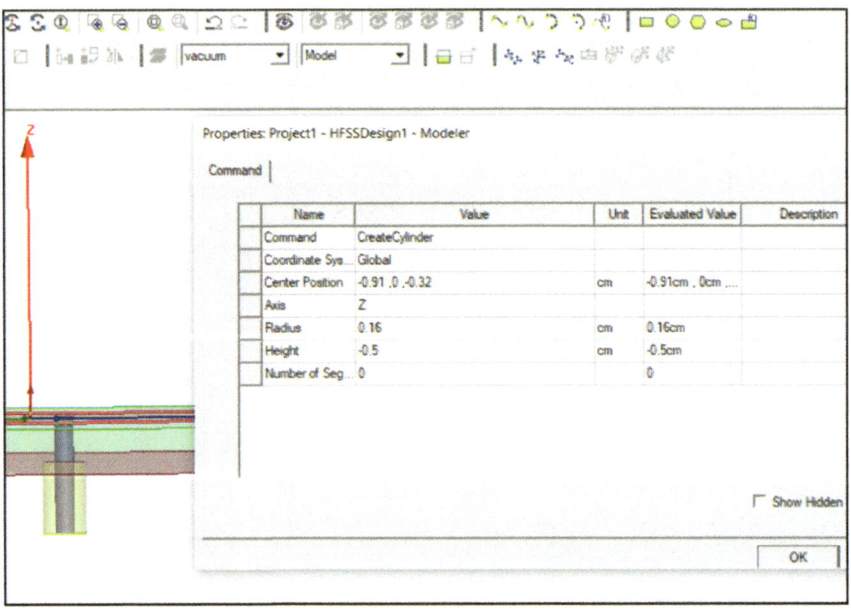
vacuum
Model
Z
Properties: Project1 - HFSSDesign1 - Modeler
Command
Name
Value
Unit
Evaluated Value
Description
Command
CreateCylinder
Coordinate Sys...
Global
Center Position
-0.91 ,0 ,-0.32
cm
-0.91cm , 0cm ,...
Axis
Z
Radius
0.16
cm
0.16cm
Height
-0.5
cm
-0.5cm
Number of Seg...
0
0
Show Hidden
OK

Fig. 5.9 .

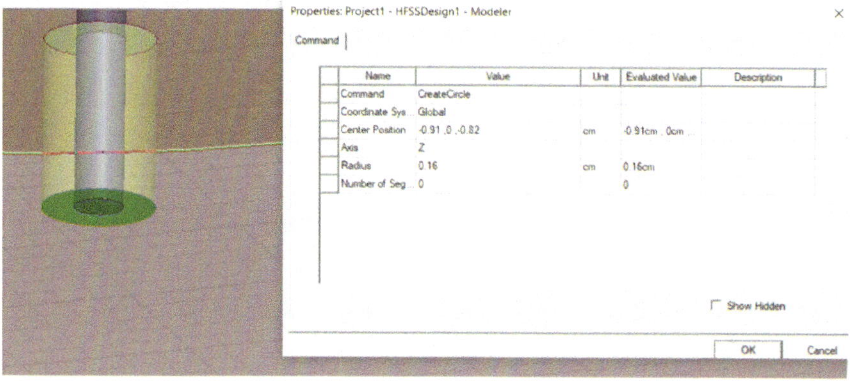

Fig. 5.10 .

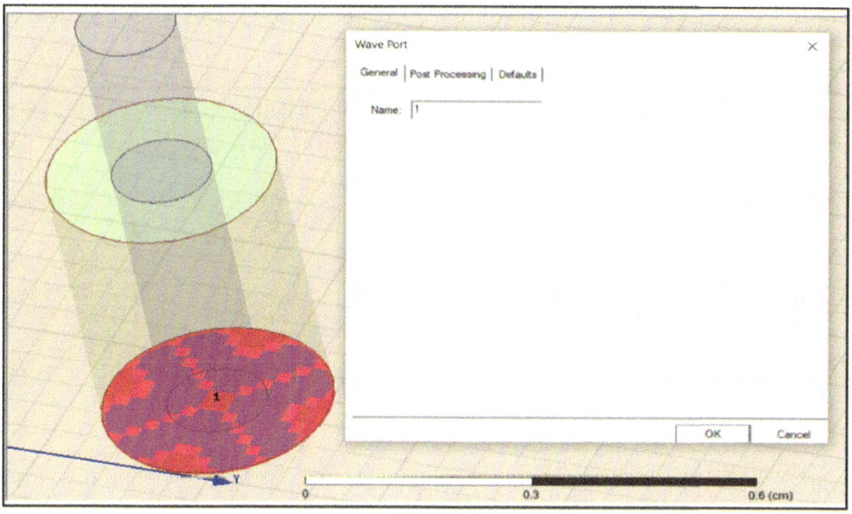

Fig. 5.11 .

15. To plot S(1,1) of the patch antenna, go to the **HFSS** toolbar > select **Result** and plot rectangular plot as shown in Fig. 5.15.
16. Now, S-parameters can be observed in rectangular results and the radiation pattern can be plotted under the far-field results as shown in Fig. 5.16.

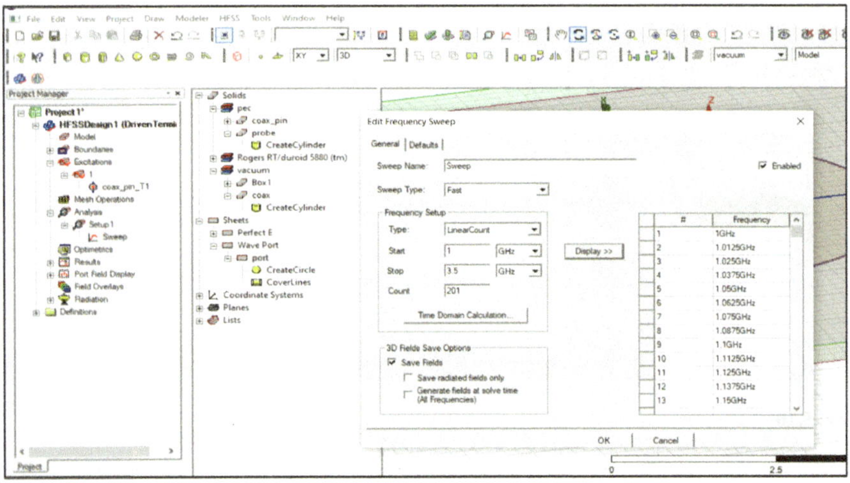

Fig. 5.12 .

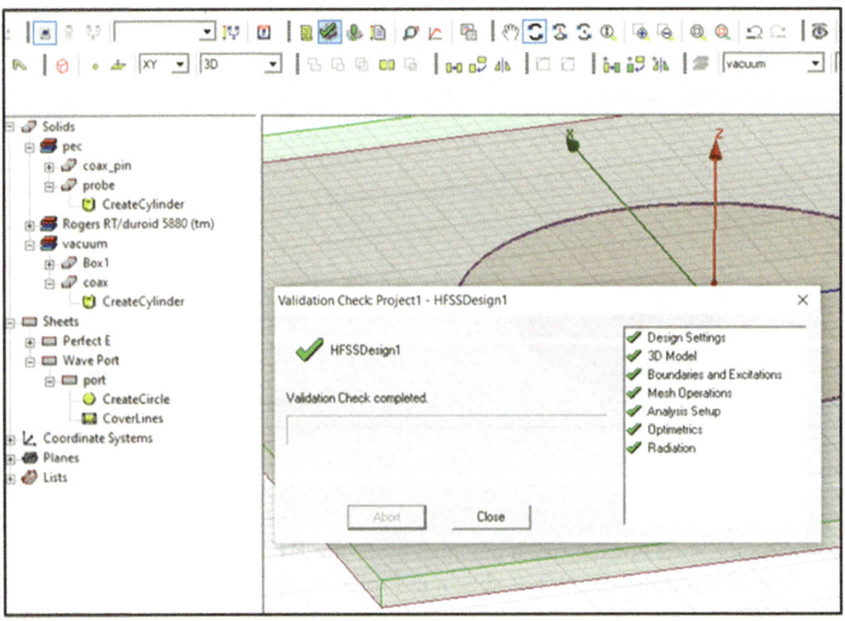

Fig. 5.13 .

5.3 COMSOL

1. From the file menu, select New, and open the **Model Wizard** as shown in Fig. 5.17.
2. Select **3D model** for the planar antenna design, as shown in Fig. 5.18.

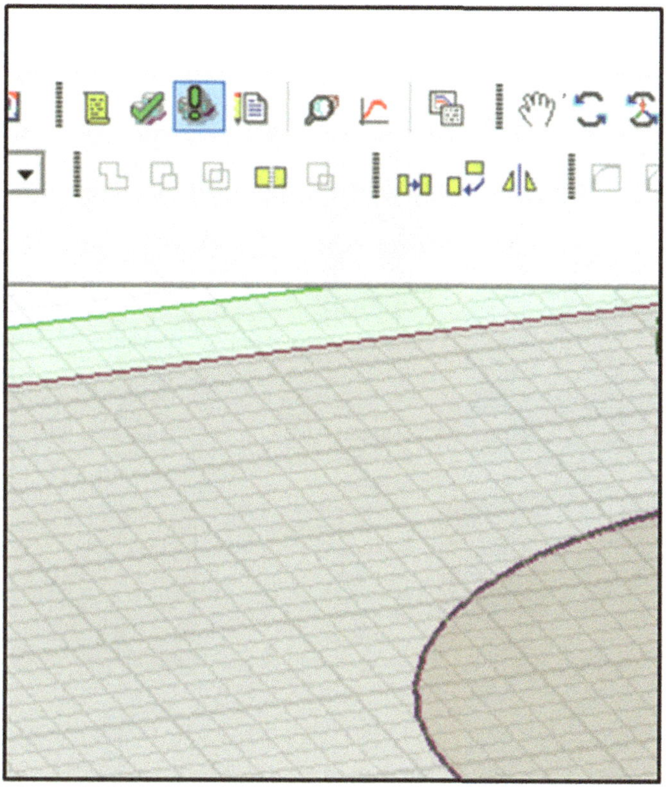

Fig. 5.14 .

3. To add physics to the project, click **Radio Frequency** ≫ **Electromagnetic Waves, Frequency Domain (emw)** as shown in Fig. 5.19.
4. Select **Dependent variable** and choose **Study** (Fig. 5.20).
5. In the study tree, select the **Frequency Domain** (Fig. 5.21).
6. In the Model Builder window, click on **Study** ≫ right-click on the **Frequency Domain** ≫ settings. In the frequencies field, type **1.575 GHz** (Fig. 5.22).
7. In global definitions, click on the **Parameters,** and add the following variables as shown in Fig. 5.23.
8. To define the units under component 1, click on the geometry and select **mm** (Fig. 5.24).
9. To define substrate, select the block from the **Geometry** toolbar and use the shown size and shape as shown in Fig. 5.25.
10. After entering all the variables, click on the **Build Selected**. A substrate will be created in the graphics window (Fig. 5.26).
11. To create a patch, from the **Geometry** toolbar, add a block of shown dimensions, and click on **Build All Objects** (Fig. 5.27).

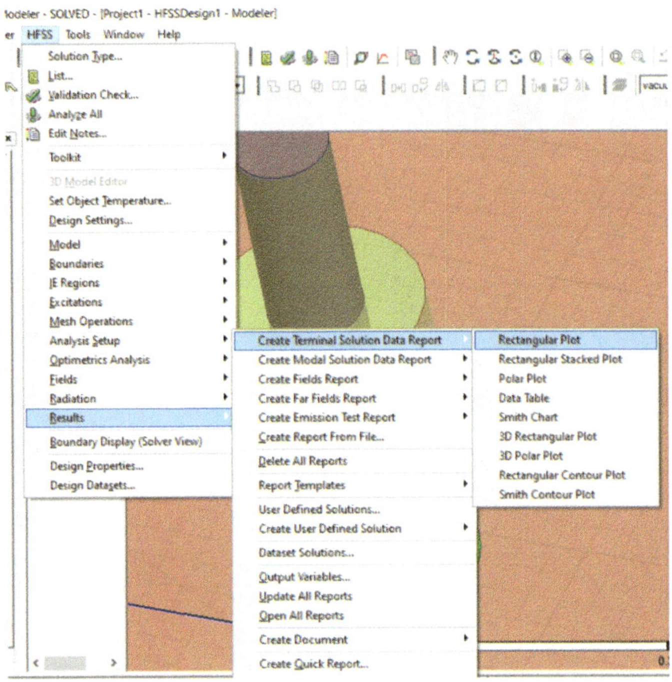

Fig. 5.15 .

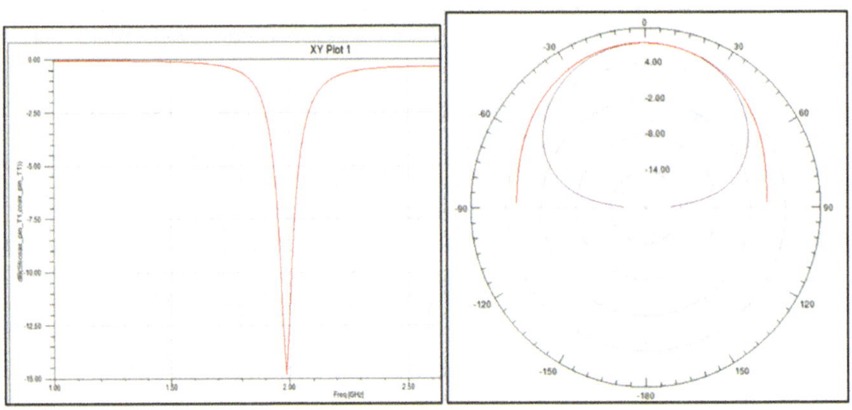

Fig. 5.16 .

12. For inset feeding, a stub is created, choose a block, add the following dimensions, and again click on **Build All Objects** (Fig. 5.28).
13. To create a mirror image of the first stub, click on the **Transforms** in the **Geometry** toolbar and choose copy (Fig. 5.29).

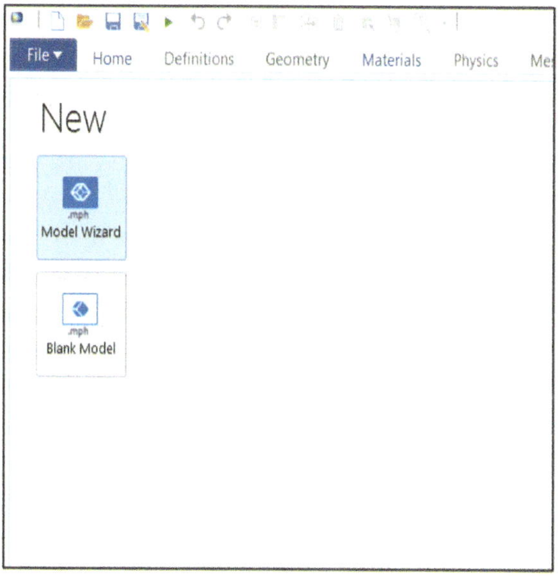

Fig. 5.17 .

14. Under **Model Builder**, click on copy, and select block 3. Under displacement bar, define x as **-w_stub-w_line** (Fig. 5.30).
15. Choose the option Difference from the **Booleans and Partitions** drop-down menu in the **Geometry** toolbar (Fig. 5.31).
16. In the difference tree, in the object to add > select block 2 component, and in objects to subtract > choose block 3 and copy 1 component. Now click on **Build All Objects** (Fig. 5.32).
17. For outer environment, to add a sphere, choose a sphere from the geometry toolbar. Add radius as l_sub, and in layer 1 as l_sub/5. Click on **Build all objects** (Fig. 5.33).
18. Select perfectly matched layer from the **Definitions** toolbar (Fig. 5.34).
19. Right-click on the **Perfectly Matched Layer 1** ≫ select **1–4** and **8–11** in the domain selection (Fig. 5.35).
20. The **View option**, which is present in the **Model Builder** window under **Component 1**, can cover the outer boundaries and helps to view interior objects. Right-click on the view and choose **Hide for Physics** (Fig. 5.36).
21. Select domain **2** and **9** in the geometry entity level (Fig. 5.37).
22. Again, right-click on the view and add **Hide for Physics 2**. Add domain **10** and **33** in the geometry entity selection (Fig. 5.38).
23. Under the **Model Builder**, right-click on the **Electromagnetic Waves, Frequency Domain (emw)**, and select **Perfect Electric Conductor**. Select boundaries **15, 20, 21** (Fig. 5.39).

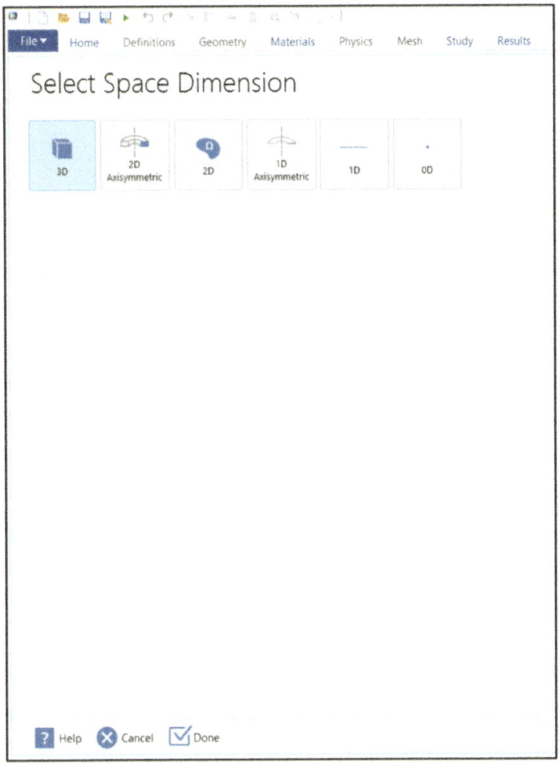

Fig. 5.18 .

24. Under the **Model Builder**, right-click on the **Electromagnetic Waves, Frequency Domain (emw)**, and select **Lumped Port** (Fig. 5.40).
25. Click on the zoom-in button and choose boundary **26** (Fig. 5.41).
26. Add material: Right-click on the materials under the components and choose add material from the library (Fig. 5.42).
27. Add **Air** from the **build in materials** (Fig. 5.43).
28. Again, add material for the substrate, and add **Blank Material** by clicking on the **Materials** (Fig. 5.44).
29. In settings pop-up, add material content as shown in Fig. 5.45.
30. To add mesh to the design, right-click on the **Mesh** under **Model Builder** navigation tree and select **Build All** (Fig. 5.46).
31. Simulate the project by clicking on the **Compute** (in the **Home** toolbar) (Fig. 5.47).
32. In the electric field toolbar, click **Evaluate Along Normal** and plot **Arrow Volume** by clicking **Plot** option in the settings (Fig. 5.48).
33. In the **Model Builder**, under the electric field node, click on **Multislice** and choose **Thermal** setting in coloring and style (Fig. 5.49).

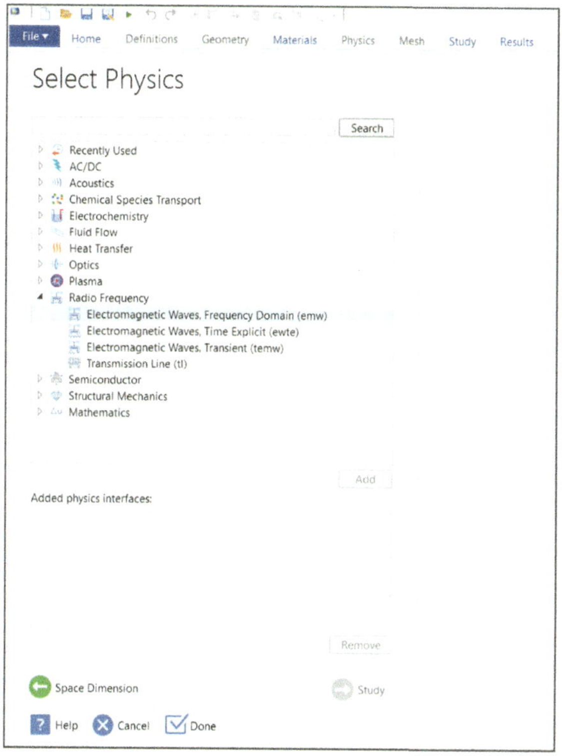

Fig. 5.19 .

34. Figure 5.50 shows the S-parameters of the design
35. Under the **Result** ≫ right-click on **2D far-field** and **Radiation Pattern 1**. Click the **Plot** option in the settings (Fig. 5.51).
36. Figure 5.52 shows the S-parameter table with the working frequency.

5.4 Computational Simulation Tool (CST) Studio

It is one of the best tools for designing high-frequency structures and electromagnetic analysis. It provides reliable and accurate results for devices such as couplers, filters, and antennas.

1. To create a new project, open the **CST software** and a project window will appear as shown; then, select **MW & RF Opticals** (Fig. 5.53).
2. Now select the work field for designing microstrip antennas through double click (Fig. 5.54).
3. A window will appear that shows different types of antennas, which are available for designing. Select **Planar** antennas as shown in Fig. 5.55.

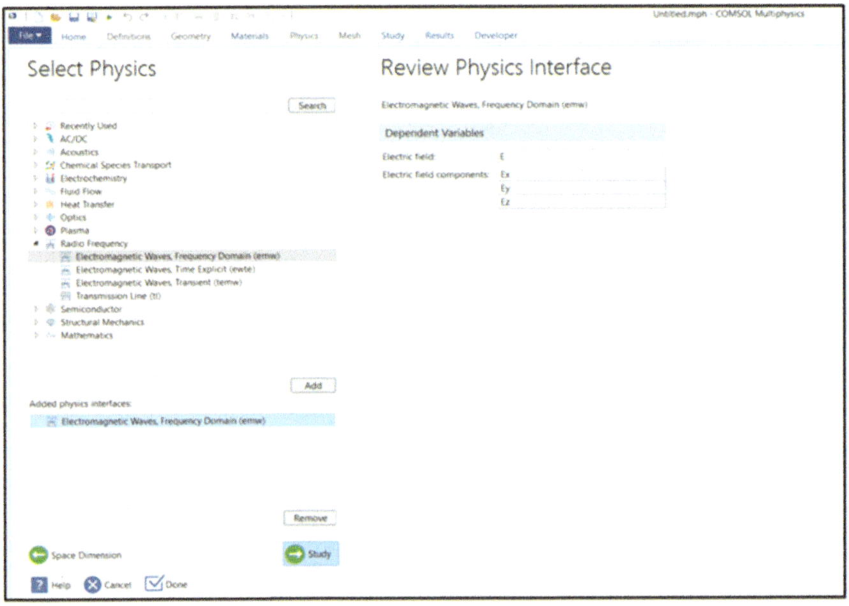

Fig. 5.20 .

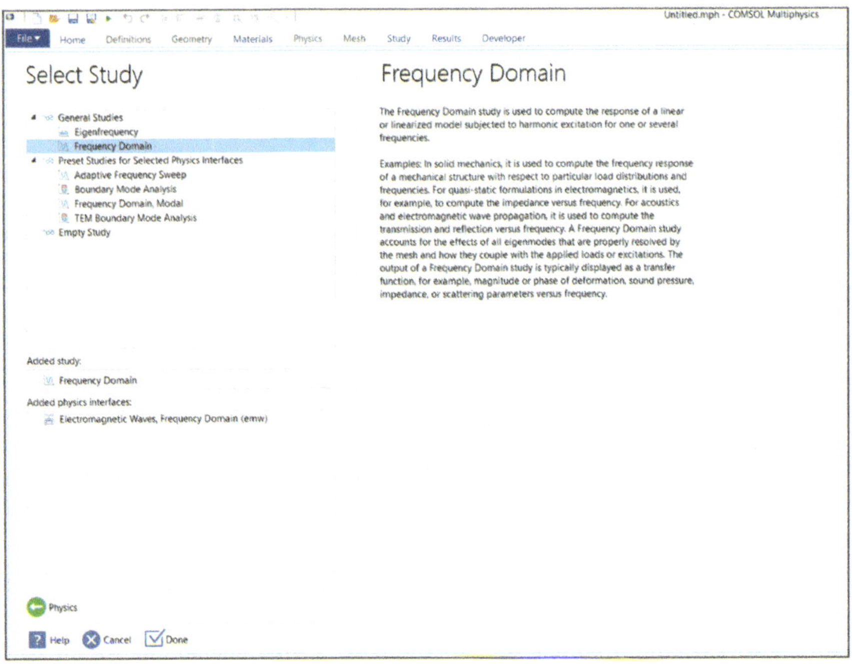

Fig. 5.21 .

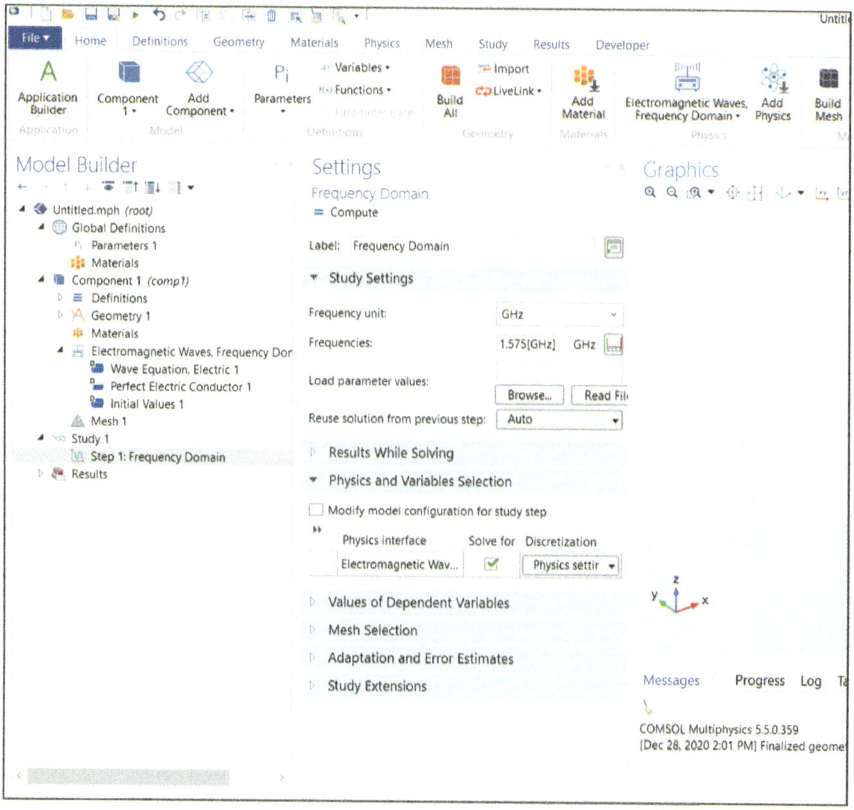

Fig. 5.22 .

4. Select the work domain to work; here, **Time Domain** is chosen for the antenna designing (Fig. 5.56).
5. After clicking next, a new project window will appear. Save this project with some name, go to **File** ≫ **Save As** ≫ and enter the name of the project (Fig. 5.57).
6. To create a substrate, click on **Modelling** and choose **Brick** as shown in Fig. 5.58.
7. A parameter window will appear, where dimensions are defined, $X_{min} = -sbx/2$ and $X_{max} = sbx/2$ and $Y_{min} = -sbY/2$ and Y_{max} = sbY/2 ($sbx = sby =$ 80 mm) (Fig. 5.59).
8. Double click on the material and choose **FR-4 (lossy)** (Fig. 5.60).
9. To define a ground plane, draw the brick, and a brick window will pop up. Define the dimensions of the substrate and material as copper as shown in Fig. 5.61.

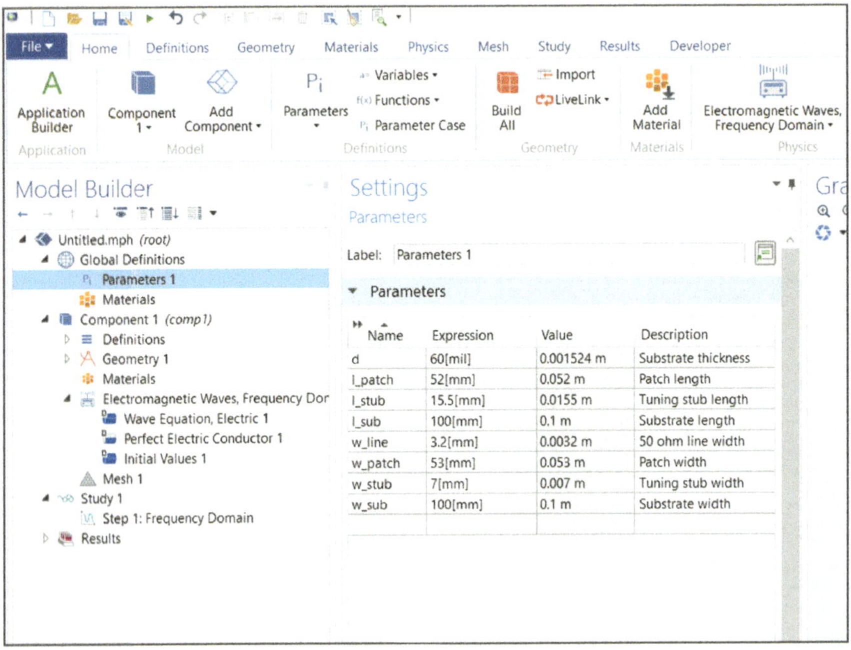

Fig. 5.23 .

10. To define a patch, draw a brick, and add dimensions to it, $X_{\min} = -ax/2$, $X_{\max} = ax/2$, $Y_{\min} = -ay/2$, and $Y_{\max} = ay/2$ ($ax = 47$ mm and $ay = 30.2$ mm) as shown in Fig. 5.62.
11. To add an inset feed, draw the **Brick**, and add dimensions to it, $X_{\min} = -lx/2$, $X_{\max} = lx/2 + \text{ins}x$, $Y_{\min} = -ay/2$, and $Y_{\max} = ay/2 + \text{ins}y$($lx = 2.98$ mm, $\text{ins}x = 1.5$ mm, $\text{ins}y = 7.16$ mm, $ay = 30.2$ mm, and $sbh = 1.5$ mm) as shown in Fig. 5.63.
12. Select the antenna patch, and click on the **Boolean Operation** ≫ **Subtract**, and chose inset as shown in Figs. 5.64 and 5.65.
13. To create an inset feeding line, draw **Brick** and add dimensions to it as shown in Fig. 5.66.
14. Merge the feedline to the patch with the help of Boolean tool as shown in Fig. 5.67.
15. To assign a port, select the **Y-axis**, and enter dimensions of the wave port as shown in Fig. 5.68.
16. Add frequency sweep by clicking on the **Frequency** in the **Simulation** tab (Fig. 5.69).
17. Check that the time domain solver is selected for simulation (Fig. 5.70).
18. Start the simulation by clicking on the start icon in the **Home** tab (Fig. 5.71).

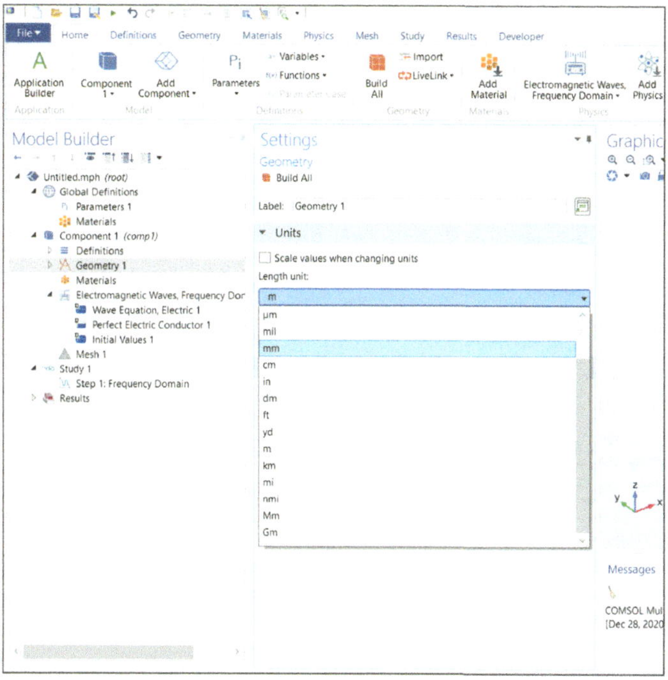

Fig. 5.24 .

19. To check S-parameters, go to the navigation tree at the leftmost side and click on **1-D results**, and choose **S-parameters**. In a similar way, one can also check 2-D far-field results (Fig. 5.72).

5.5 Implementation of Impedance and Rectifier Circuit on the Advanced Design System (ADS)

Steps to create a schematic design:

1. In the main window workspace, click on the file, and then go to the new option.
2. In a new tab, you will find multiple options for creating a new workspace, library, schematic, layout, symbol, substrate, and hierarchy. Go to the new schematic.
3. The window will appear as shown in Fig. 5.73. Write the name of the schematic cell (e.g., rectifier).
4. The new schematic window is created, as shown in Fig. 5.74.
5. Design of rectifier circuit.

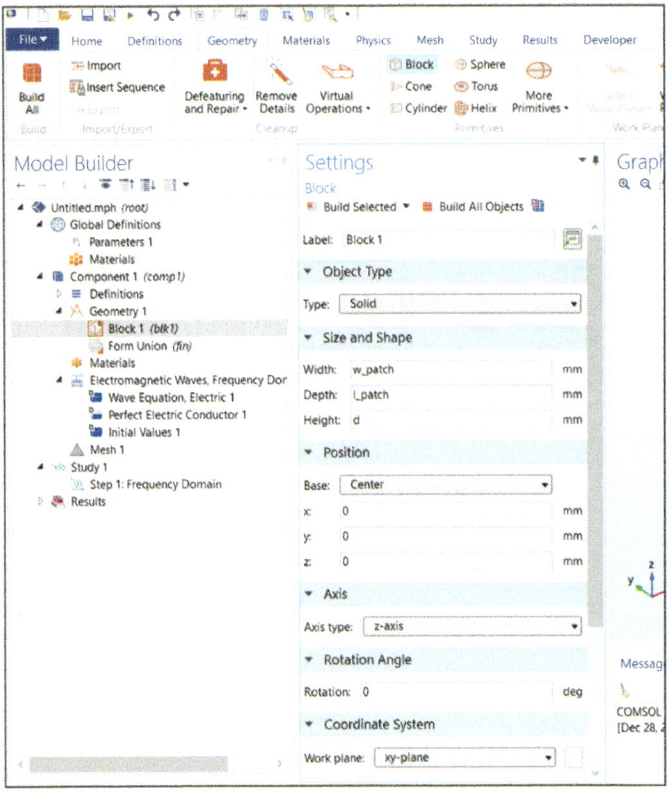

Fig. 5.25 .

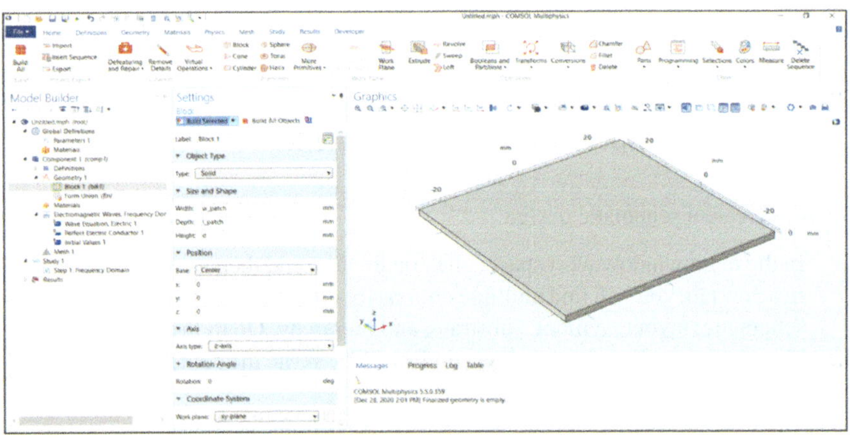

Fig. 5.26 .

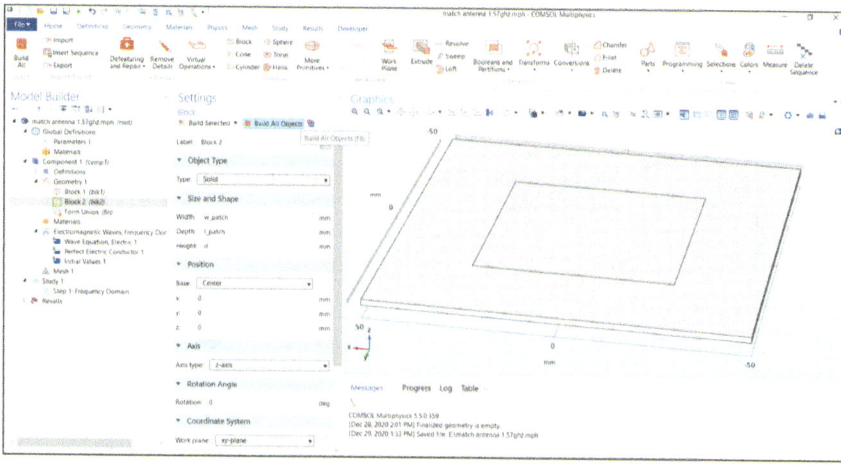

Fig. 5.27 .

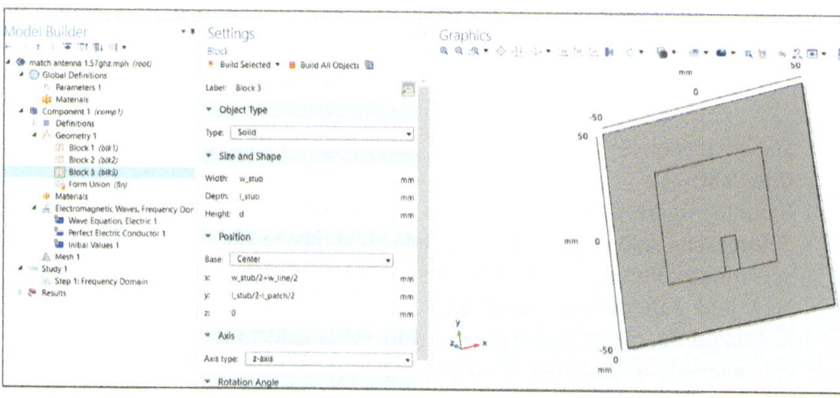

Fig. 5.28 .

(a) Start with the power source from the component palette, shown at the leftmost side of the schematic in Fig. 5.75. There are various sources available, and select source frequency domain, as we want to design the circuit topology in the frequency domain.

(b) From the left palette, click on the **Pn_tone** to design a dual-frequency band circuit. And, to design a single-frequency band, **P1_tone** is used (Fig. 5.76).

(c) Double click on the **Pn_tone** component, and a window will appear as shown in Fig. 5.77.

(d) Add two operating frequencies and two power sources for the two frequencies. Click on the apply button and then OK (Fig. 5.78).

(e) Now add lumped components from the left side palette, as shown in Fig. 5.79.

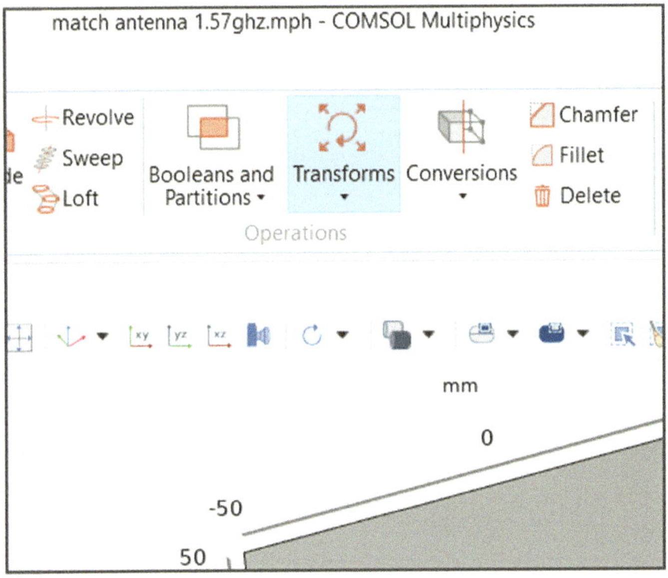

Fig. 5.29 .

(f) Select inductors and capacitors according to the circuit requirements (Fig. 5.80).
(g) From the figure, select transmission line for the stub designing (Fig. 5.81).
(h) Select radial stub and transmission line, as shown in Fig. 5.82, and place them accordingly. Further, select inductors and capacitors also, as shown in Fig. 5.83.
(i) Now add a diode from the library palette, here a voltage doubler is used, and therefore, two diodes are employed. Also, select the diode model and the characteristic of the diode from the datasheet as shown in Fig. 5.84.
(j) Complete the circuit with the load capacitor and the load resistor as shown in Figs. 5.85 and 5.86.
(k) Harmonic simulator: For the simulation of nonlinear circuits in the frequency domain, harmonic balance simulator is the most suitable technique for microwave and RF problems.

(i) Select the **HB simulation** from the library palette (Fig. 5.87).
(ii) Place the harmonic balance simulation component, and to add the frequency and sweep parameter, double click on the **HB component** as shown in Figs. 5.88, 5.89, and 5.90.
(iii) Click on the **Simulation** icon, and once the simulation is completed, a result window will appear as shown in Figs. 5.91 and 5.92.
(iv) Click on the rectangular plot and select **Vout** to plot the output DC voltage with respect to the input power (Fig. 5.93).
(v) Select the **DC component** and click **OK** for the graph as shown in Fig. 5.94.
(vi) A plot of DC output voltage is shown in Fig. 5.95.

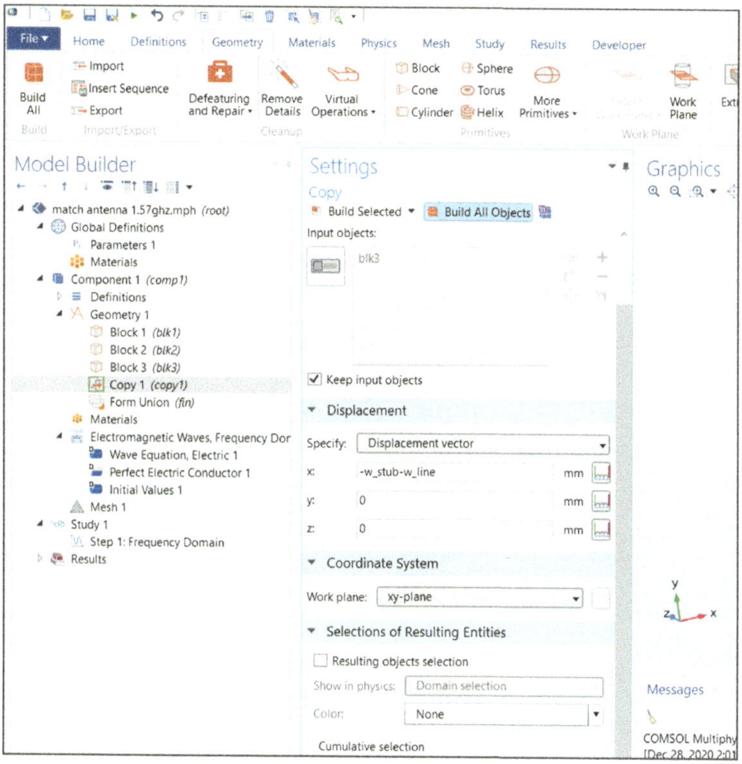

Fig. 5.30 .

6. Add S-parameter simulation.

(i) To select the **S-parameter component**, choose the simulation S-parameter from the left side as shown in Fig. 5.96.
(ii) Double click on the **S-parameter controller** in the schematic window to add the start and stop frequencies with the step size.
(iii) Again, simulate by clicking on the simulation icon. A result window will appear, and select S (1,1) in dB as shown in Fig. 5.97.
(iv) Figure 5.98 shows the S-parameters of the rectifier.

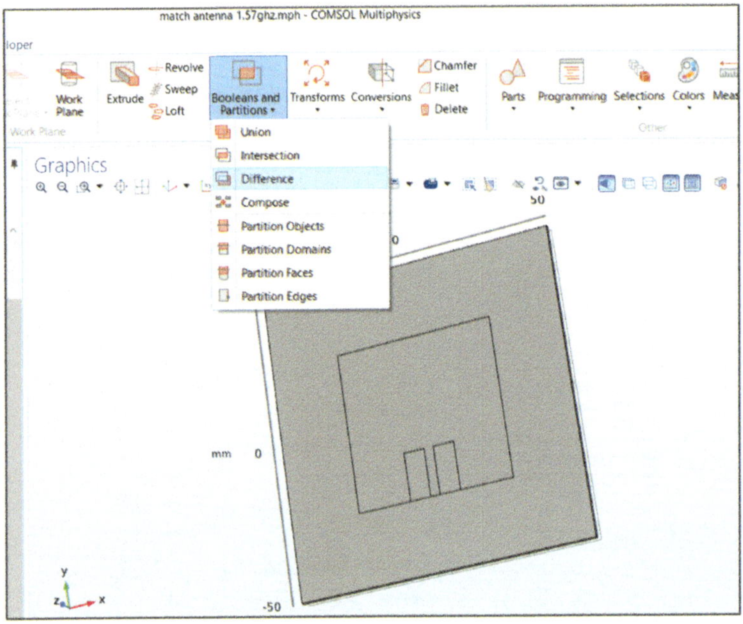

Fig. 5.31 .

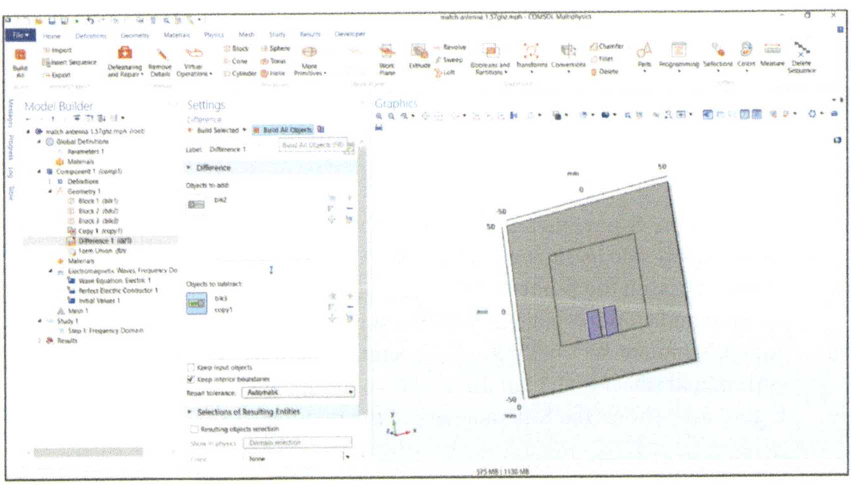

Fig. 5.32 .

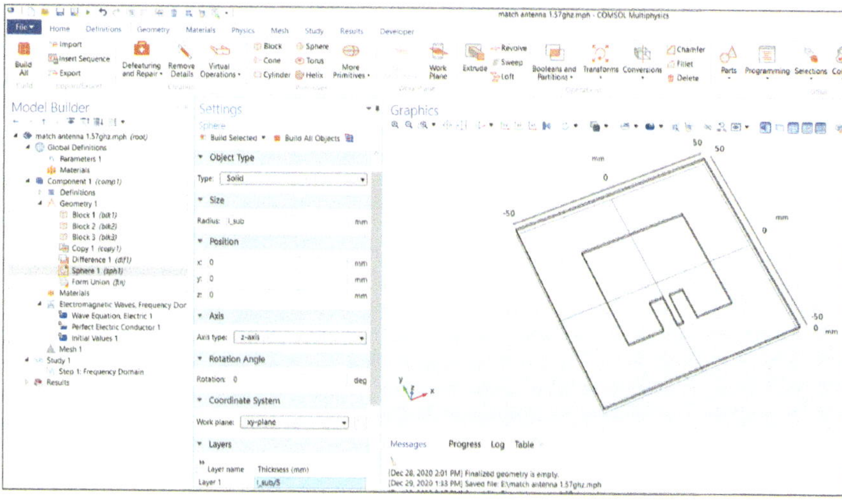

Fig. 5.33 .

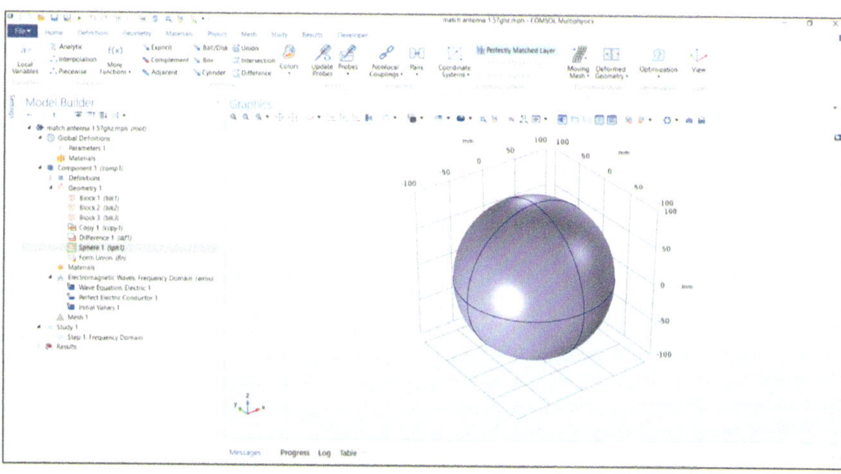

Fig. 5.34 .

5.6 Integration of the Antenna with a Rectifier (Co-simulation)

The antenna is designed using a different tool, and the schematic circuit of the rectifier is designed on the ADS tool. The process of combining the rectifier and antenna is called integration or co-simulation. The co-simulation can be performed by using ADS.

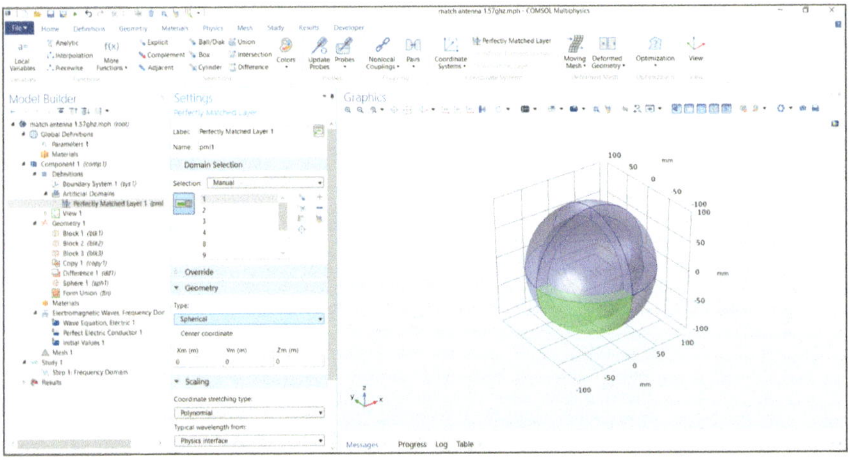

Fig. 5.35 .

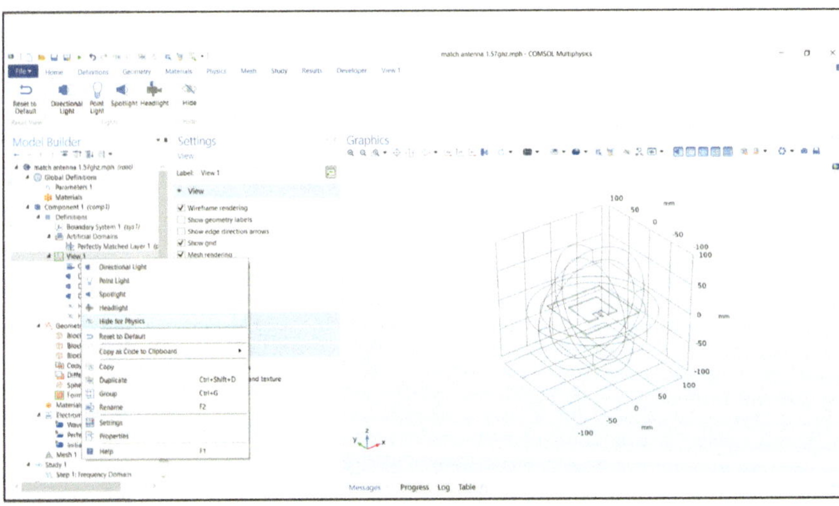

Fig. 5.36 .

1. Exporting the results of the antenna:

(i) Open the antenna design in the HFSS as shown in Fig. 5.99.
(ii) After validation and simulation, click on the solution data icon as shown in Fig. 5.100.
(iii) Check **Display all freqs**.
(iv) Check **S-matrix** for S-parameters and **Z-matrix** for Z-parameters, and click on the export matrix data as shown in Fig. 5.101.
(v) Save the file by adding a file name as **s.Np** (Fig. 5.102).
(vi) The figure shows the value of impedance chosen is **50 Ω**, and the results are exported for an impedance of 50 Ω (Fig. 5.103).

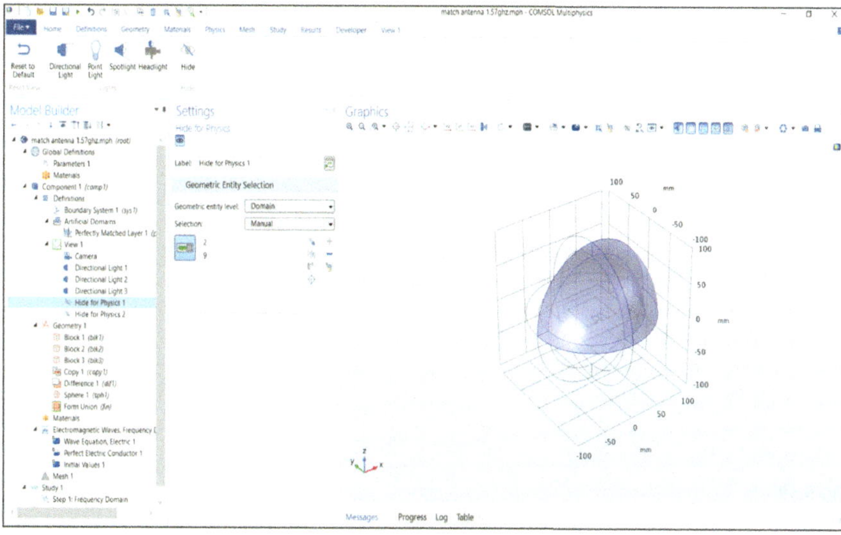

Fig. 5.37 .

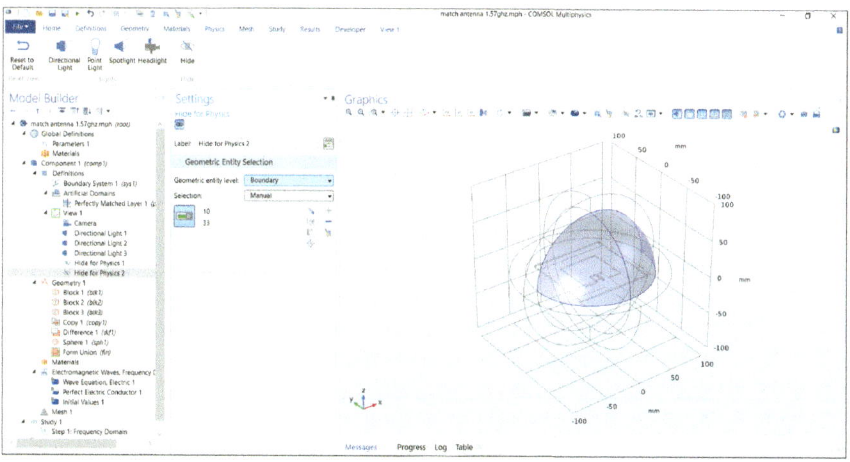

Fig. 5.38 .

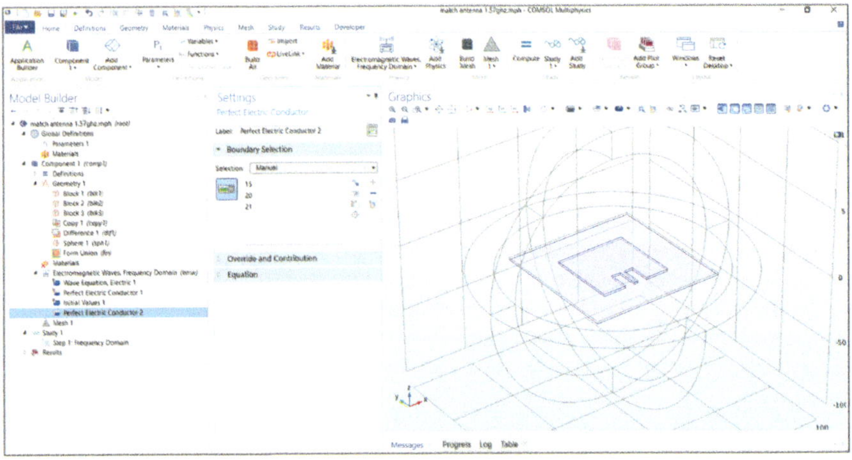

Fig. 5.39 .

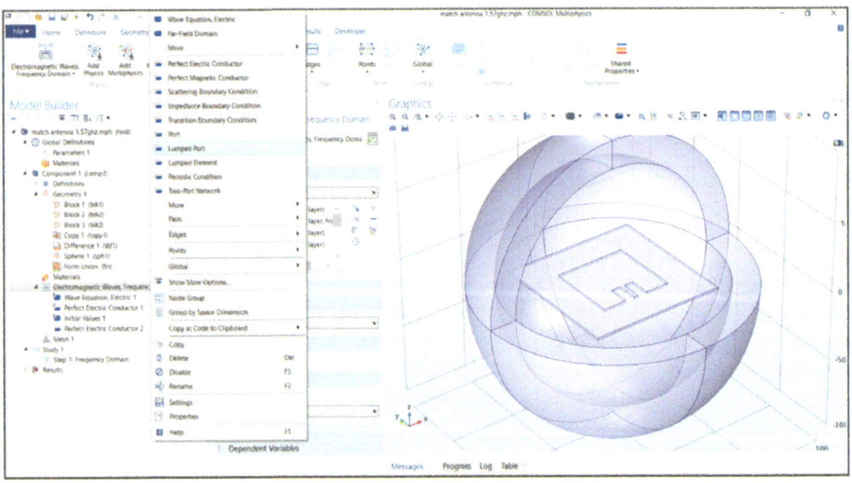

Fig. 5.40 .

5.7 Integration of HFSS and ADS

1. From the library palette, select the **Data Items** as shown in Fig. 5.104.
2. Select the **1-port S-Parameter file** component and place it (Fig. 5.105).
3. Double click on the **S1P data component** to browse the exported results of the antenna as shown in Fig. 5.106.
4. As shown in Fig. 5.107, click on the browse button and open the file exported from the HFSS.
5. Figure shows that the antenna results are successfully added to the rectifier circuit and simulate it again for the results (Fig. 5.108).

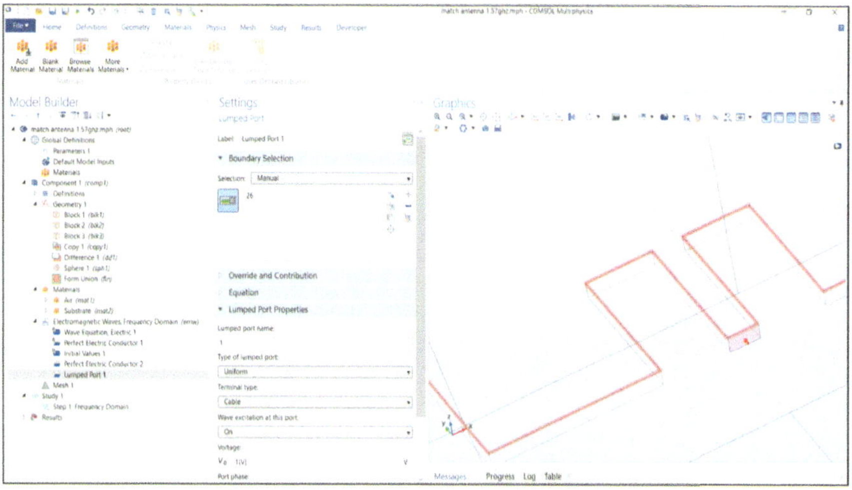

Fig. 5.41 .

5.8 Circuit Tuning and Optimization at Microwave Range

The desired results can be possible by changing the calculated values of the components. To change the value of each component and simulate every time is time taking process and sometimes complex. The software provides the tools (tuning and optimization) to vary the values.

(i) Tuning

This is a manual process, where desired results can be achieved by changing the values manually.

(a) Open the schematic window and simulate the circuit.
(b) Click on the tuning icon as shown in Fig. 5.109.

1. A tune parameter window will appear in front of the schematic window, as shown in Fig. 5.110.
2. Now, component value can be changed sequentially, as shown in the figure. Check the component and click OK (Fig. 5.111).
3. For the transmission line, the parameters, such as width, length, and angle, can be checked after clicking on the microstrip line component (e.g., MRSTUB and MLIN) as shown in Figs. 5.112 and 5.113.
4. The Tune Parameters window varies the values manually as shown in Fig. 5.114.
5. After changing some values, variations in the results can be observed (Fig. 5.115).

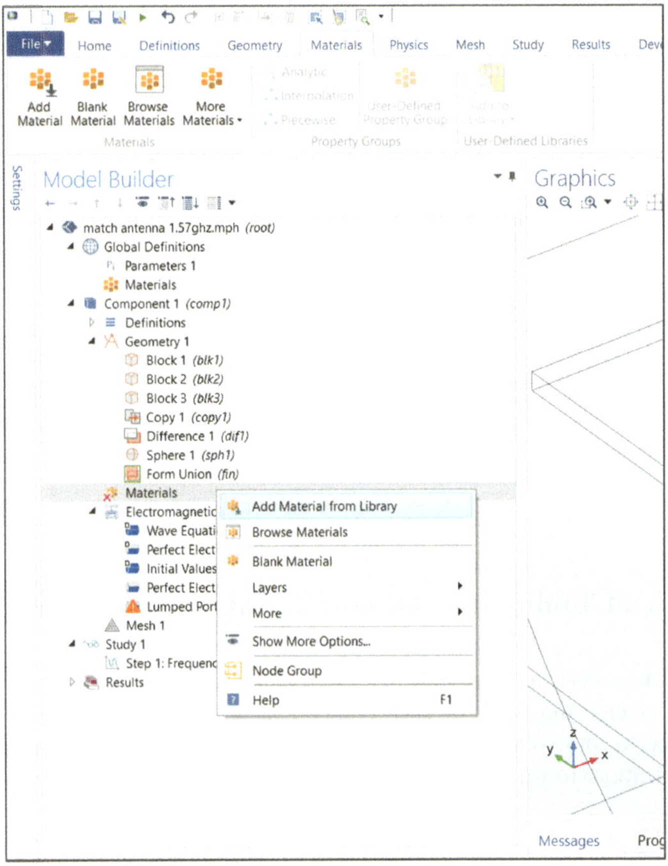

Fig. 5.42 .

(ii) Optimization

Optimization is different from tuning because it is not a manual operation. The goal is set accordingly for the desired outcome, and through the optimization process, ADS modifies the calculated values of the components. The values of the components should be within the realistic limits in such a way that it is technically feasible to accomplish the desired purpose.

Open the schematic window and select the **Optim/Stat/DOE** from the library palette as shown in Fig. 5.116.

1. From the palette, select the goal component and place the schematic window, as shown in Fig. 5.117.
2. Double click on the goal to set the parameters. Here, dB(S(1,1)) in expression and SP1 in the analysis are set as shown in Fig. 5.118.
3. As shown in the figure, edit the independent variable as frequency and click OK (Fig. 5.119).
4. Set the limit with minimum and maximum frequency range as shown in Fig. 5.120.

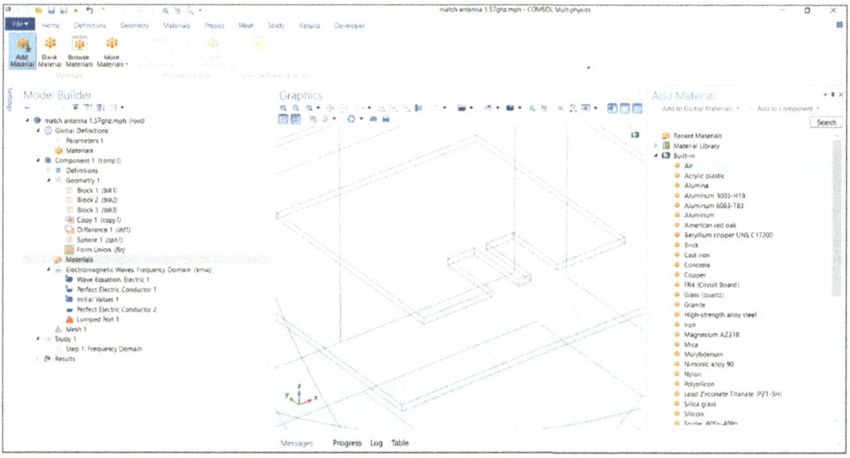

Fig. 5.43 .

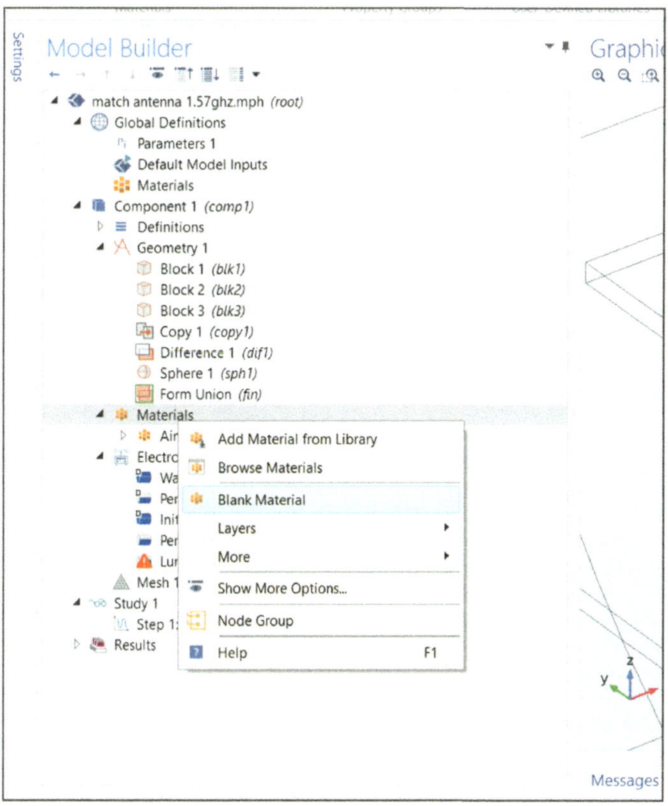

Fig. 5.44 .

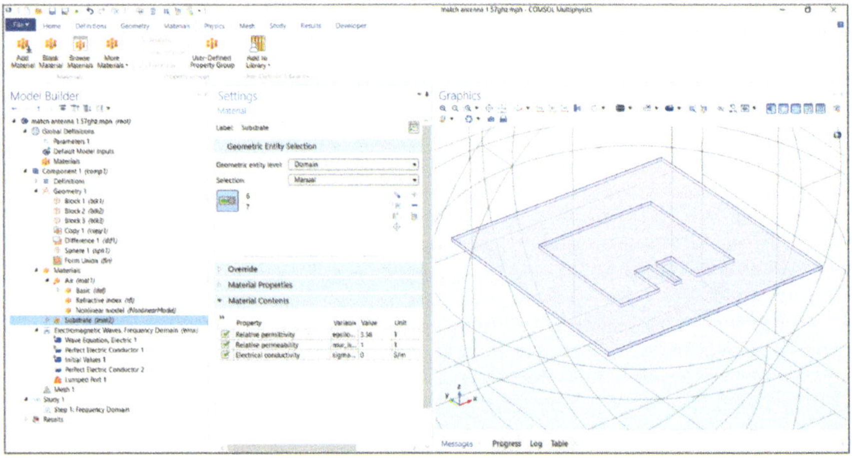

Fig. 5.45 .

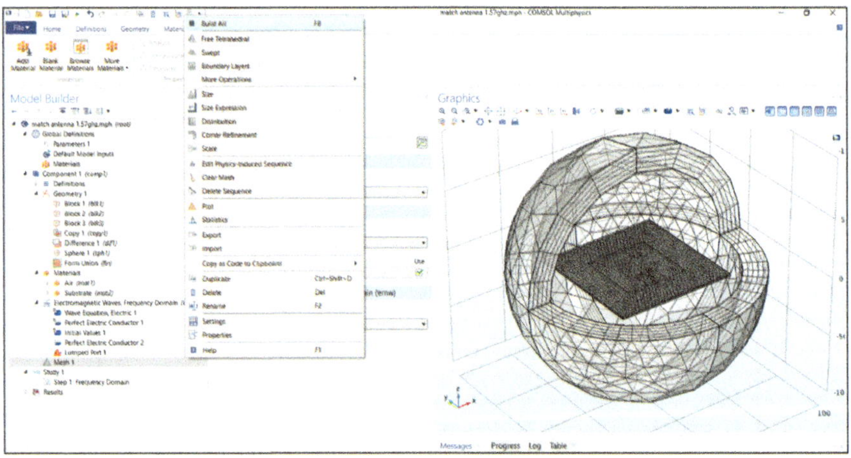

Fig. 5.46 .

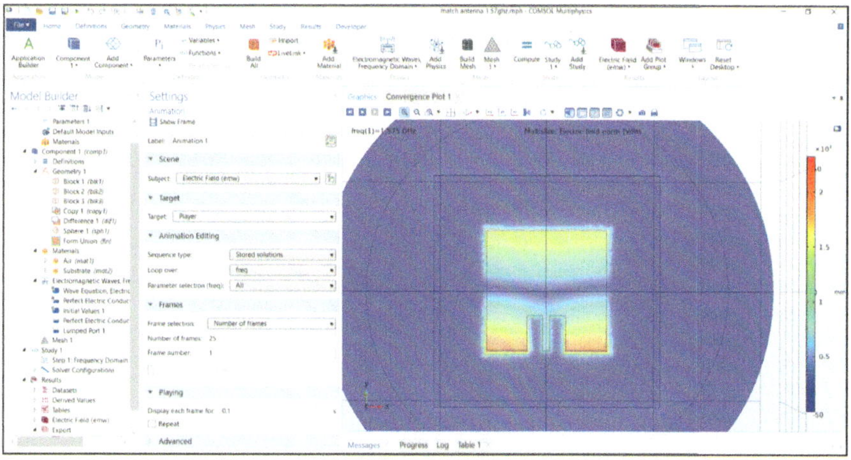

Fig. 5.47 .

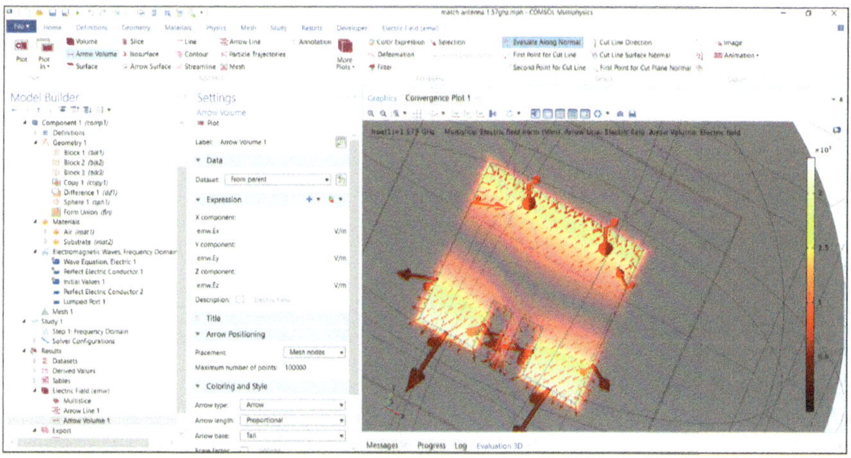

Fig. 5.48 .

Setting up optimization controller

5. After setting the goal limit, select the optimization controller with the name given optim, as shown in Fig. 5.121.
6. Double click on OPTIM and select gradient in the optimization type, as shown in Fig. 5.122.
7. Click clear all and only select optim type and Maxiters (Fig. 5.123).
8. Choose the simulate tab (Fig. 5.124).
9. Select the simulation variable setup, where the component values appear as shown in Fig. 5.125.

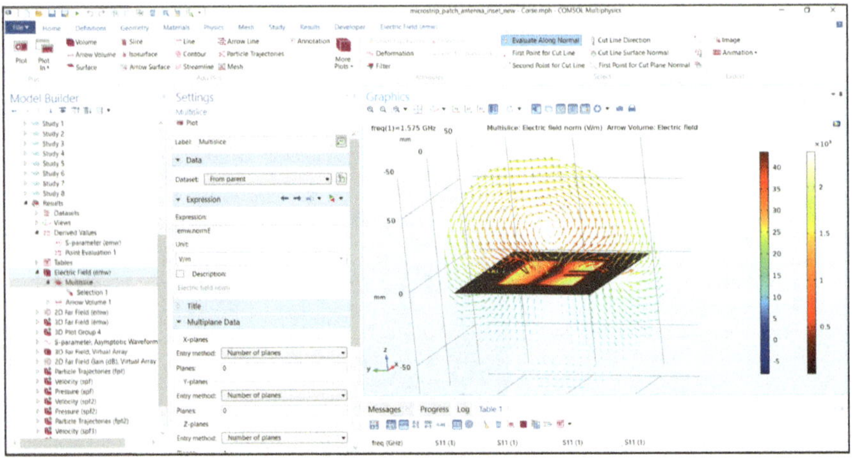

Fig. 5.49 .

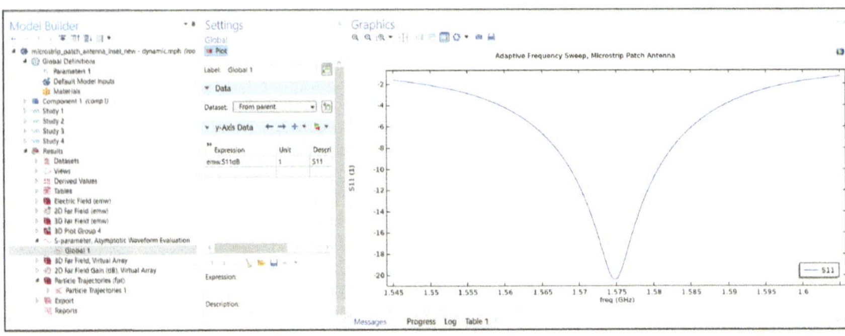

Fig. 5.50 .

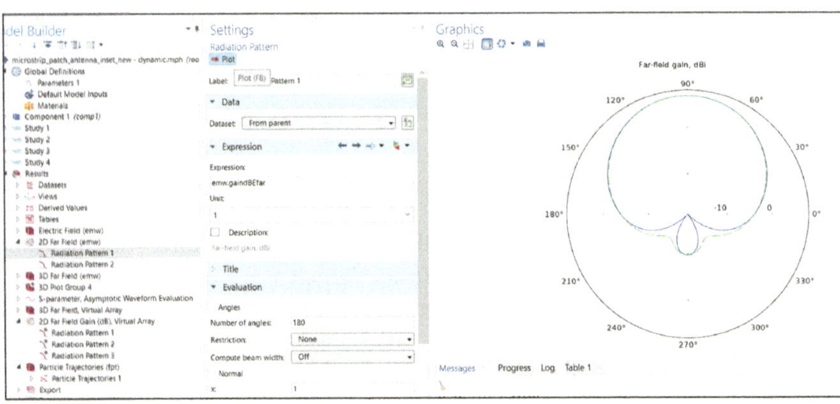

Fig. 5.51 .

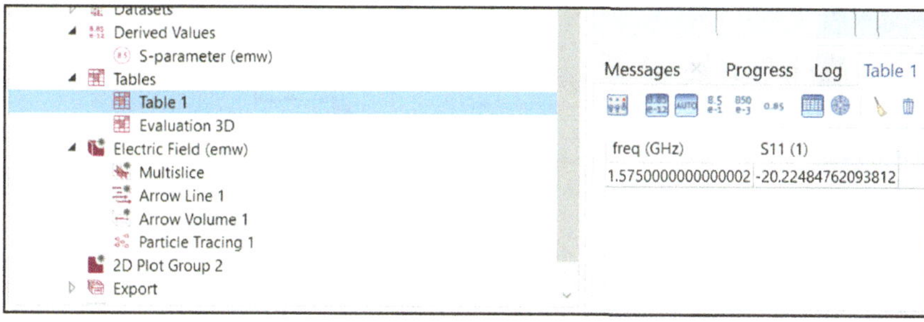

Fig. 5.52 .

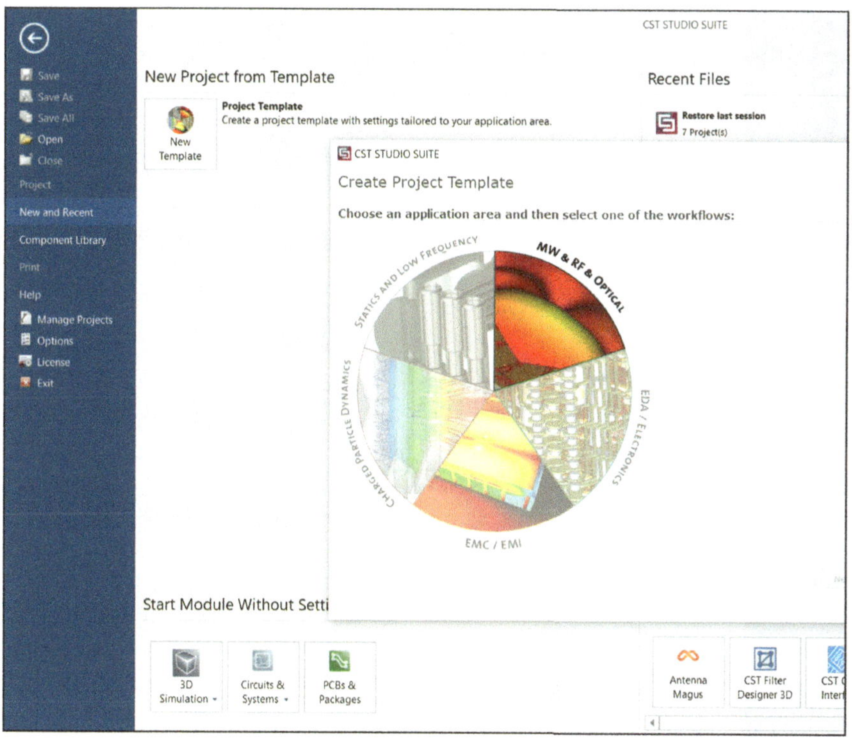

Fig. 5.53 .

10. Check all parameters for optimization (Fig. 5.126).
11. Click on optimize icon, present at the top of the schematic window, as shown in Fig. 5.127.
12. Optimization cockpit window will appear and optimize the values to achieve the desired output, as shown in Fig. 5.128.

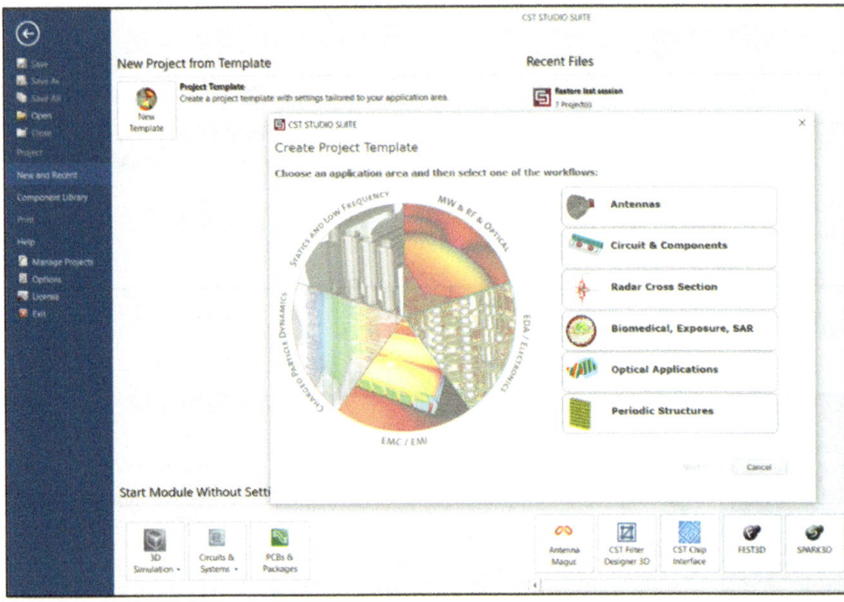

Fig. 5.54 .

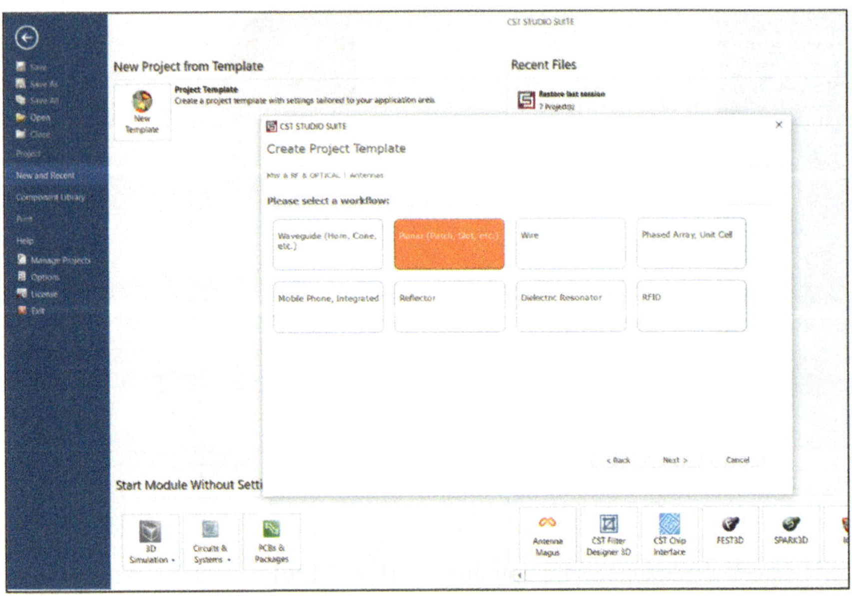

Fig. 5.55 .

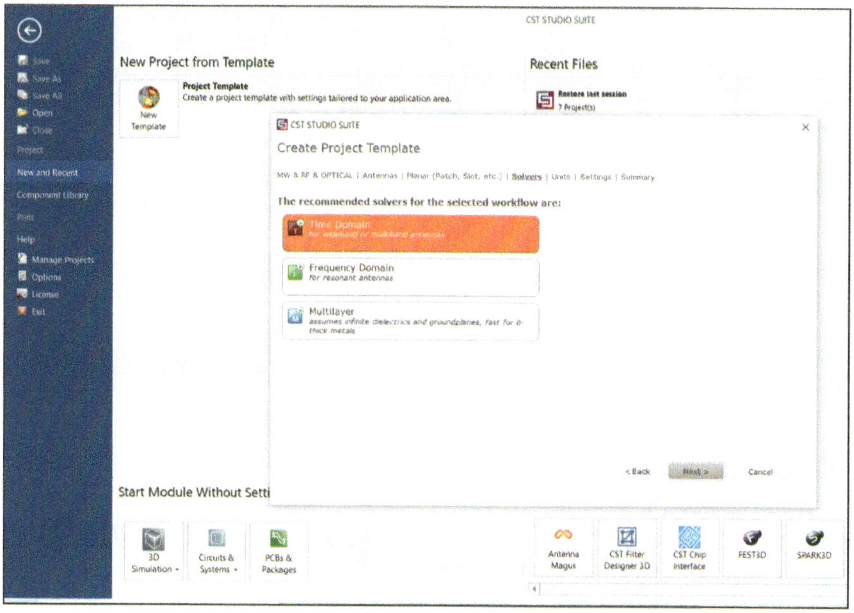

Fig. 5.56 .

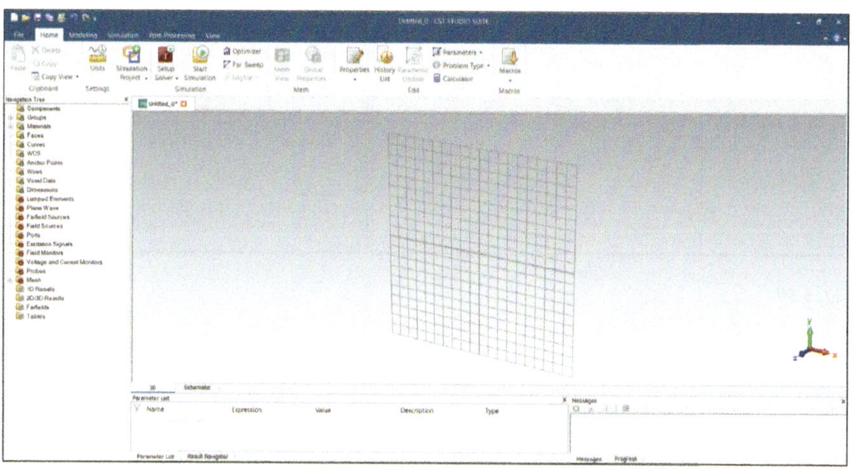

Fig. 5.57 .

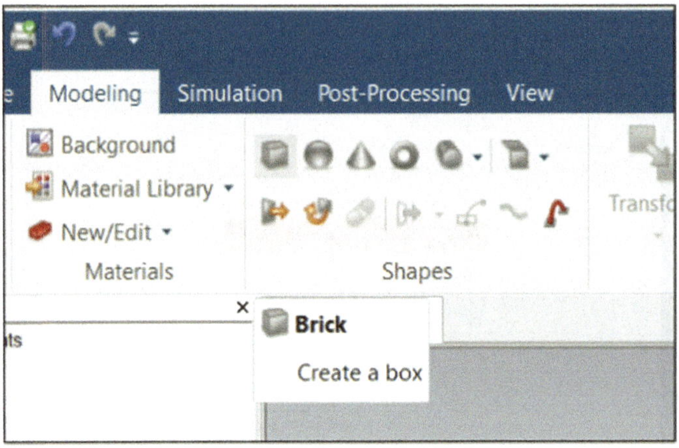

Fig. 5.58 .

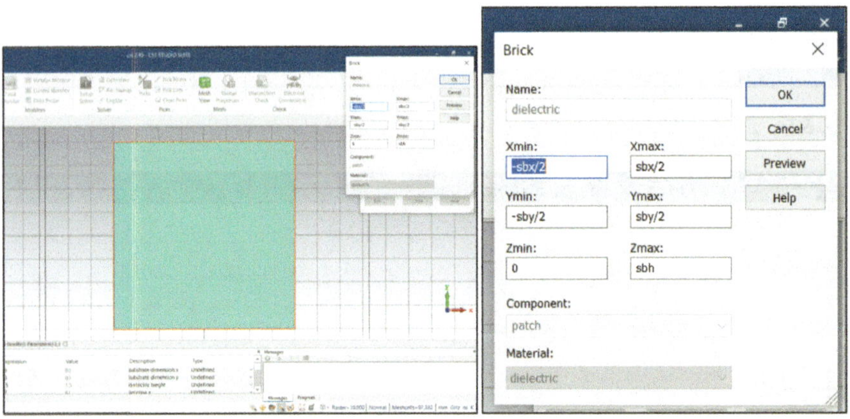

Fig. 5.59 .

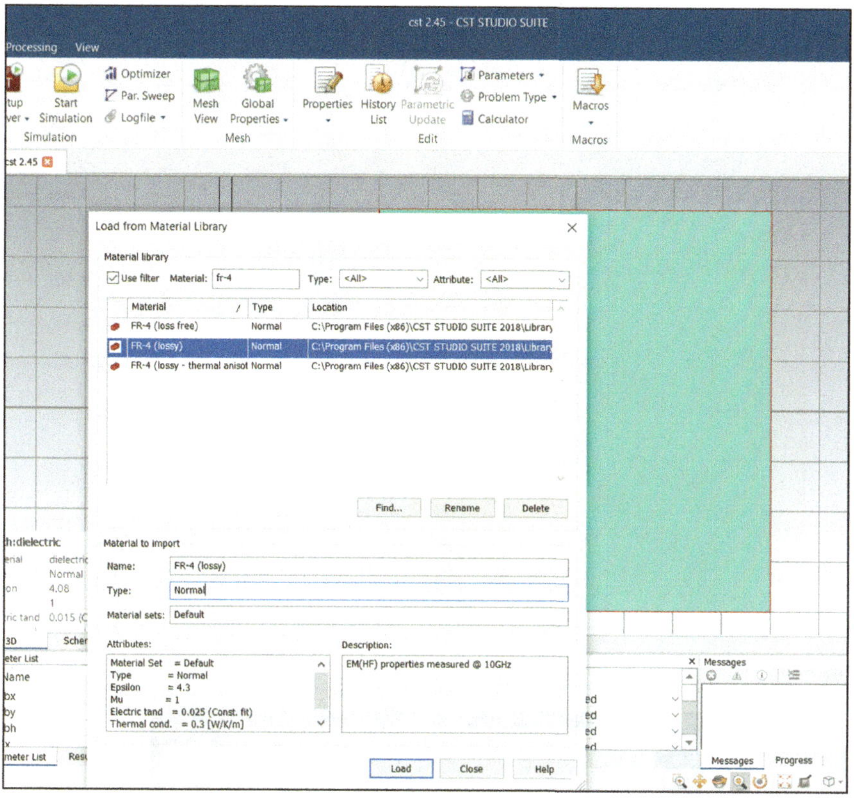
cst 2.45 - CST STUDIO SUITE
Processing
View
Optimizer
Par. Sweep
Logfile
Start Simulation
Simulation
Mesh View
Global Properties
Mesh
Properties
History List
Parametric Update
Parameters
Problem Type
Calculator
Edit
Macros
Load from Material Library
Material library
Use filter
Material: fr-4
Type: <All>
Attribute: <All>
Material
Type
Location
FR-4 (loss free) Normal C:\Program Files (x86)\CST STUDIO SUITE 2018\Library
FR-4 (lossy) Normal C:\Program Files (x86)\CST STUDIO SUITE 2018\Library
FR-4 (lossy - thermal anisot Normal C:\Program Files (x86)\CST STUDIO SUITE 2018\Library
Find...
Rename
Delete
Material to import
Name: FR-4 (lossy)
Type: Normal
Material sets: Default
Attributes:
Material Set = Default
Type = Normal
Epsilon = 4.3
Mu = 1
Electric tand = 0.025 (Const. fit)
Thermal cond. = 0.3 [W/K/m]
Description:
EM(HF) properties measured @ 10GHz
Load
Close
Help
Messages
Progress

Fig. 5.60 .

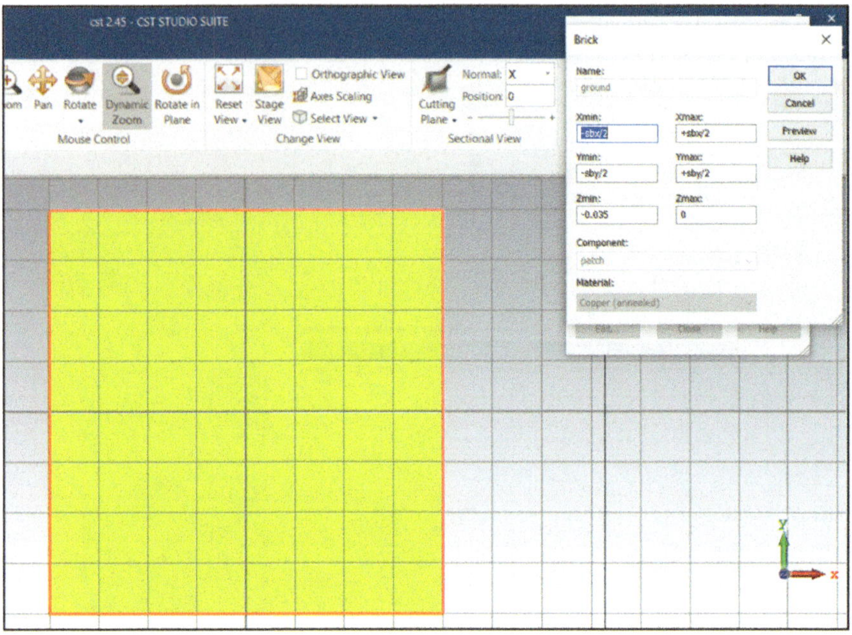

Fig. 5.61 .

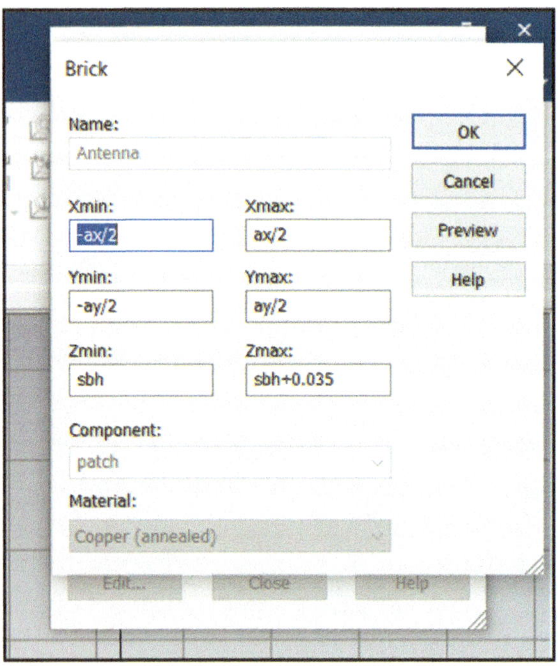

Fig. 5.62 .

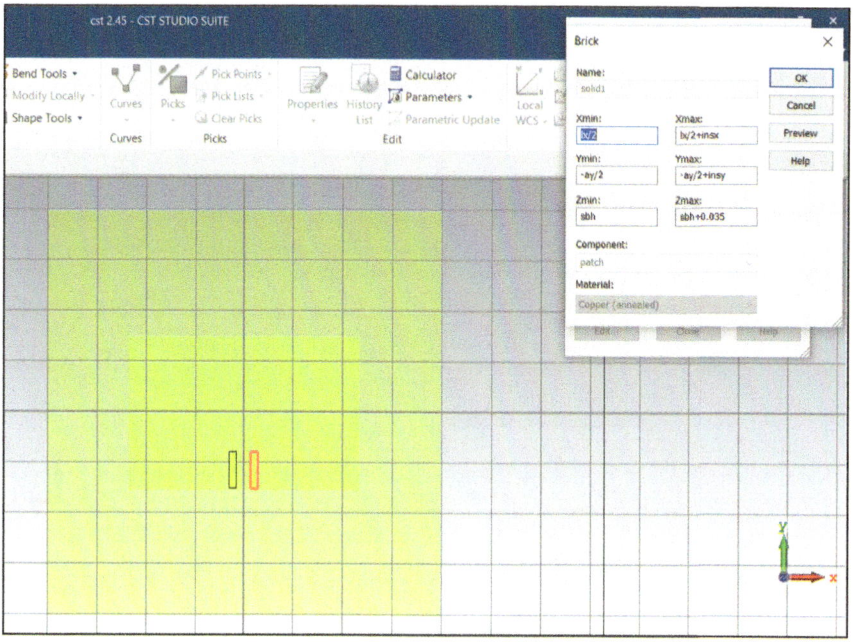

Fig. 5.63 .

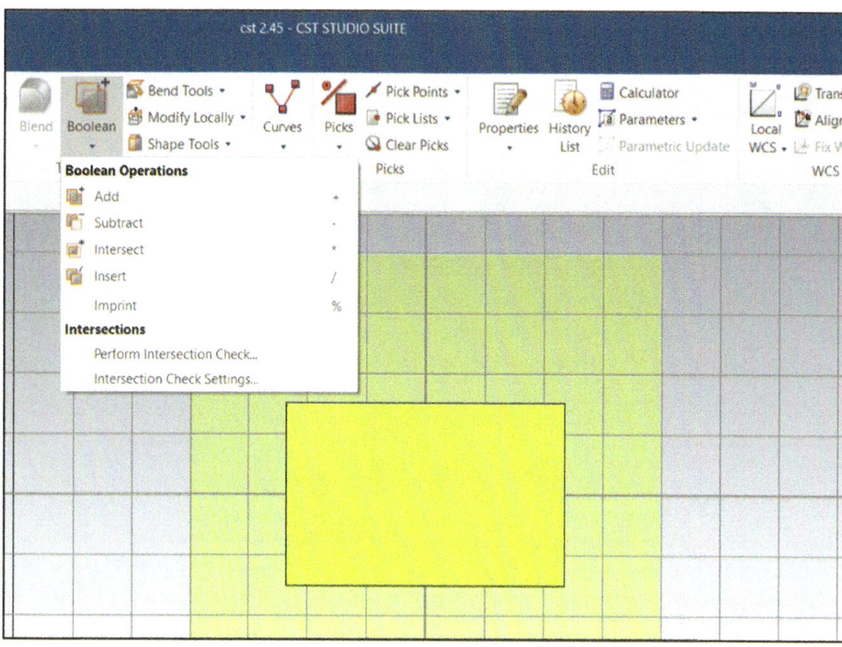

Fig. 5.64 .

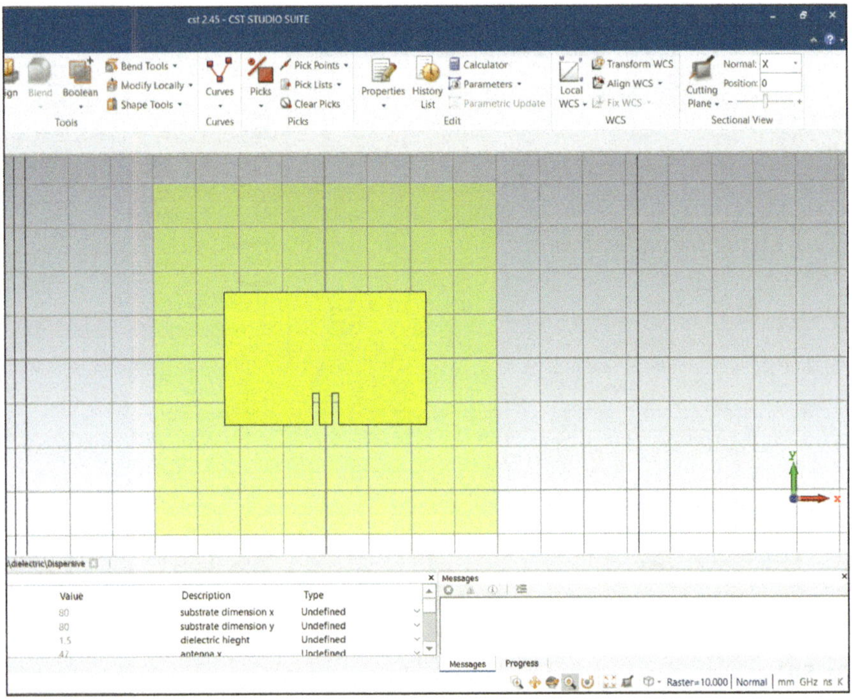
cst 2.45 - CST STUDIO SUITE
Bend Tools
Modify Locally
Shape Tools
Boolean
Tools
Curves
Picks
Pick Points
Pick Lists
Clear Picks
Properties
History List
Calculator
Parameters
Parametric Update
Edit
Local WCS
Transform WCS
Align WCS
Fix WCS
WCS
Cutting Plane
Normal: X
Position: 0
Sectional View
Value
Description
Type
80
substrate dimension x
Undefined
80
substrate dimension y
Undefined
1.5
dielectric hieght
Undefined
Messages
Progress
Raster=10.000
Normal
mm GHz ns K

Fig. 5.65 .

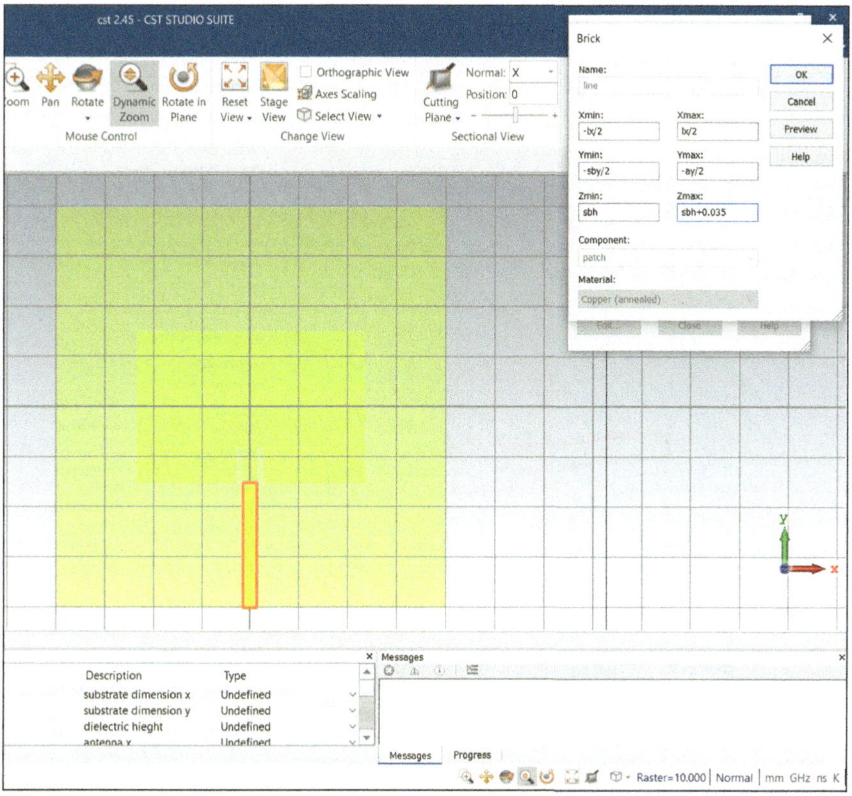

Fig. 5.66 .

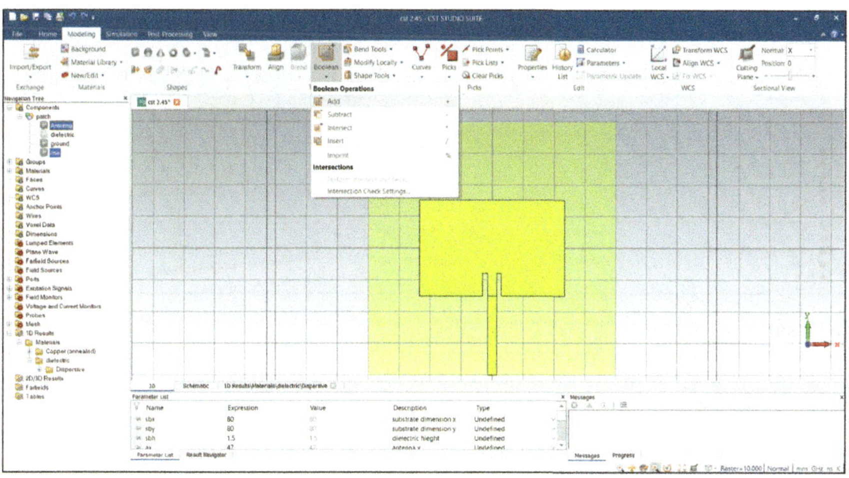

Fig. 5.67 .

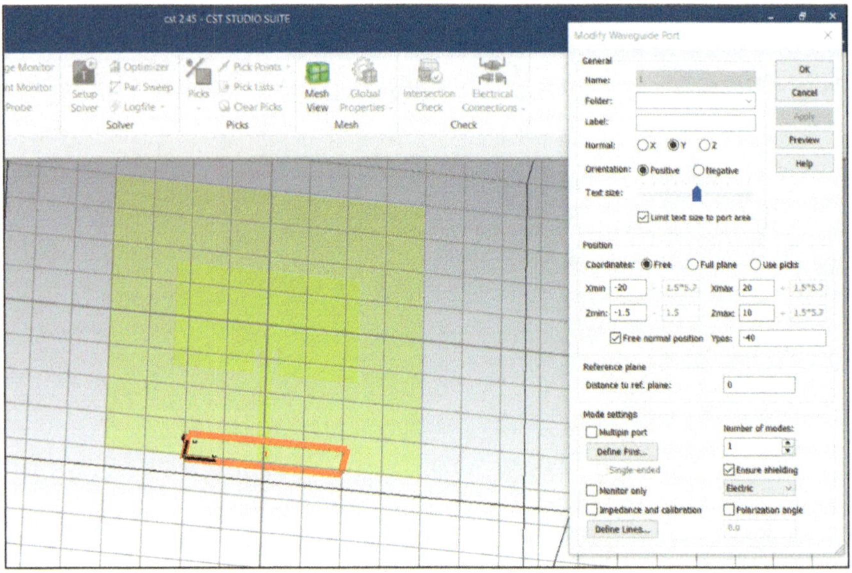

Fig. 5.68 .

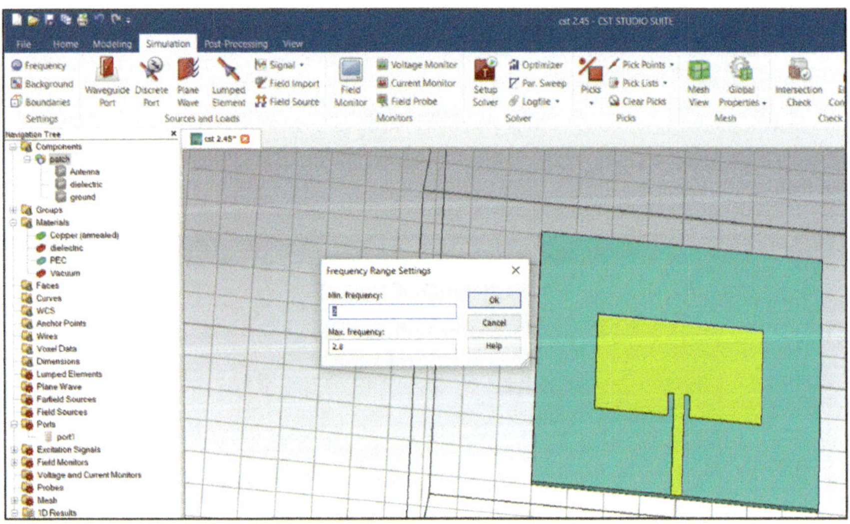

Fig. 5.69 .

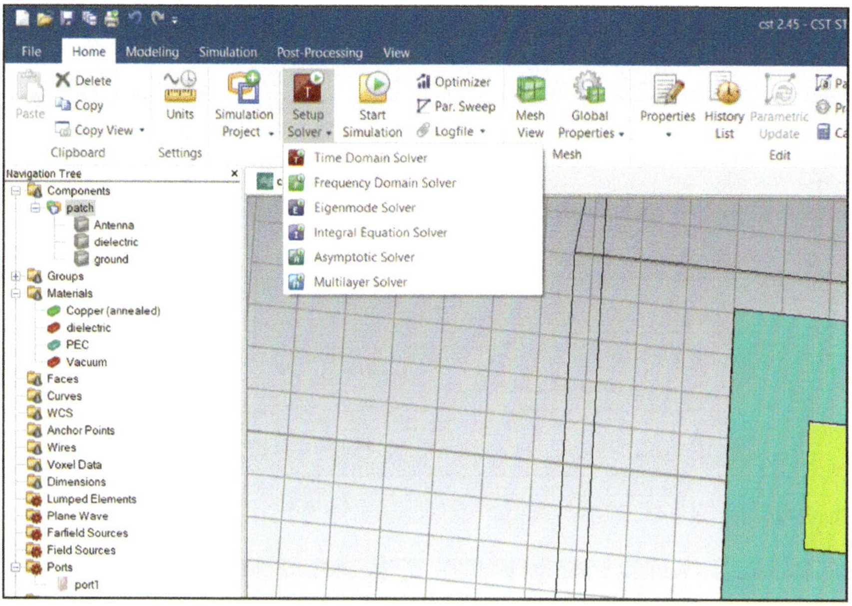

Fig. 5.70 .

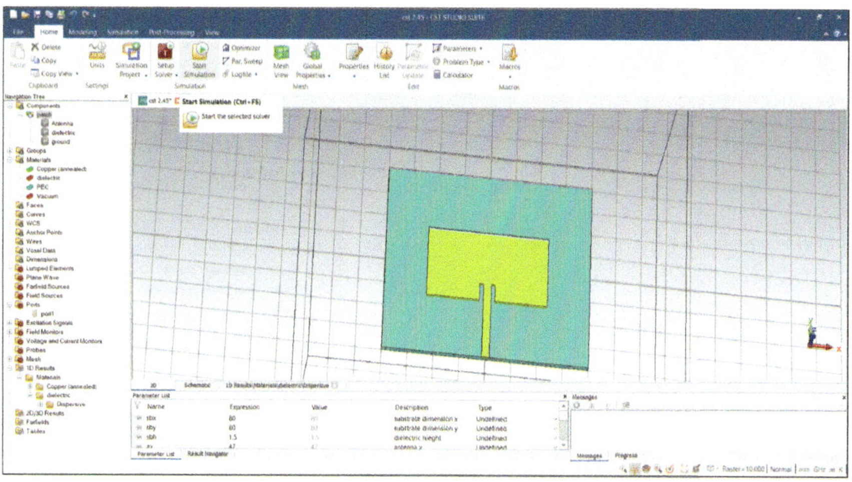

Fig. 5.71 .

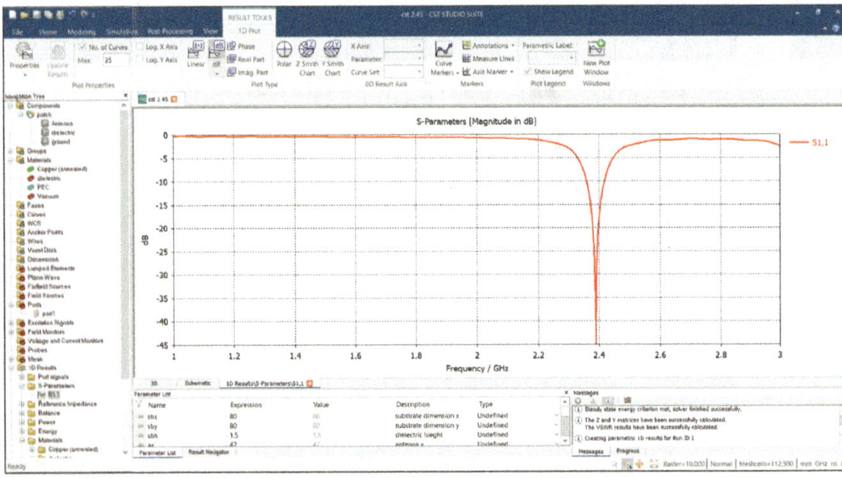

Fig. 5.72 .

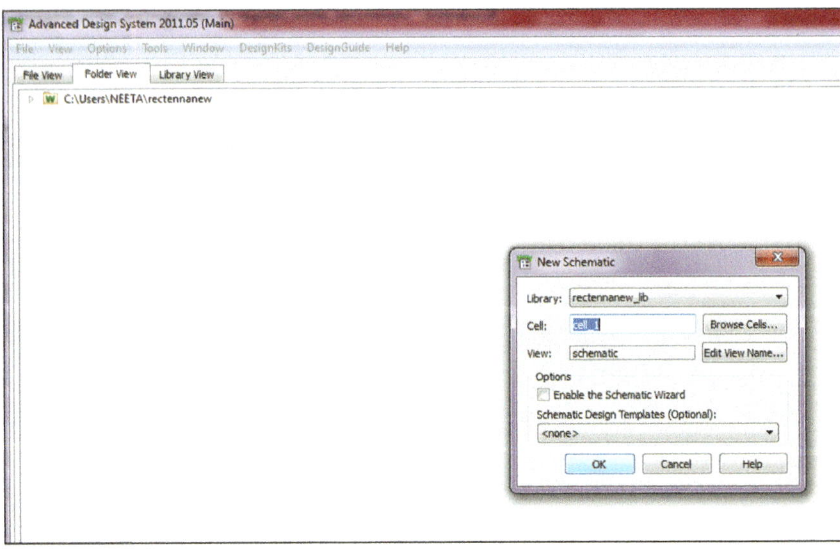

Fig. 5.73 .

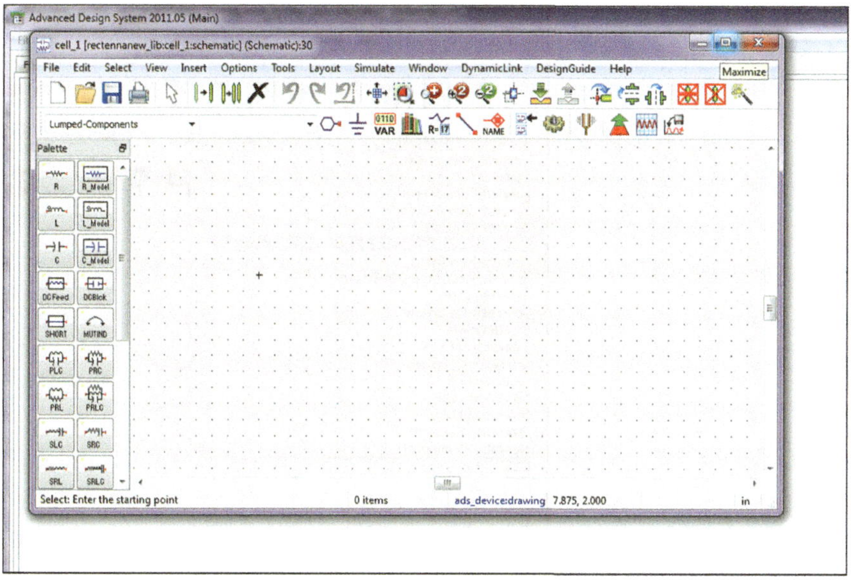
Advanced Design System 2011.05 (Main)
cell_1 [rectennanew_lib:cell_1:schematic] (Schematic):30
File Edit Select View Insert Options Tools Layout Simulate Window DynamicLink DesignGuide Help
Maximize
Lumped-Components
VAR
NAME
Palette
R
R_Model
L
L_Model
C
C_Model
DC Feed
DCBlock
SHORT
MUTIND
PLC
PRC
PRL
PRLC
SLC
SRC
SRL
SRLC
Select: Enter the starting point
0 items
ads_device:drawing
7.875, 2.000
in

Fig. 5.74 .

Fig. 5.75 .

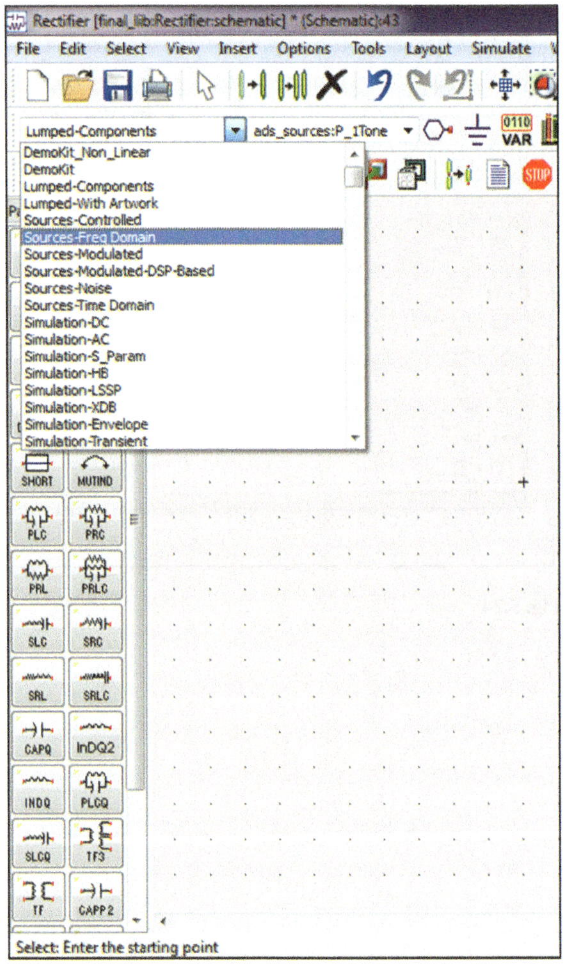
Rectifier [final_lib:Rectifier:schematic] * (Schematic):43
File Edit Select View Insert Options Tools Layout Simulate
Lumped-Components
ads_sources:P_1Tone
VAR
DemoKit_Non_Linear
DemoKit
Lumped-Components
Lumped-With Artwork
Sources-Controlled
Sources-Freq Domain
Sources-Modulated
Sources-Modulated-DSP-Based
Sources-Noise
Sources-Time Domain
Simulation-DC
Simulation-AC
Simulation-S_Param
Simulation-HB
Simulation-LSSP
Simulation-XDB
Simulation-Envelope
Simulation-Transient
SHORT
MUTIND
PLC
PRC
PRL
PRLC
SLC
SRC
SRL
SRLC
CAPQ
InDQ2
INDQ
PLCQ
SLCQ
TF3
TF
CAPP2
Select: Enter the starting point

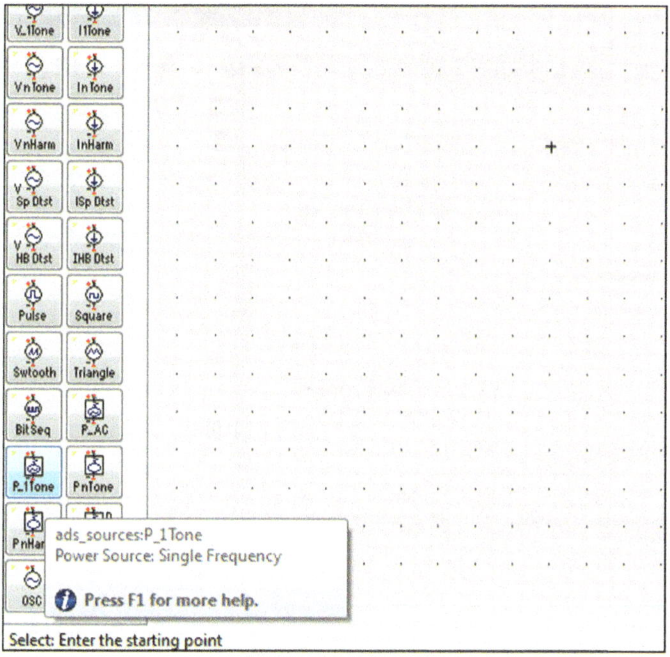
V_1Tone
I1Tone
VnTone
InTone
VnHarm
InHarm
V Sp Dtst
ISp Dtst
V HB Dtst
IHB Dtst
Pulse
Square
Swtooth
Triangle
BitSeq
P_AC
P_1Tone
PnTone
PnHar
OSC
ads_sources:P_1Tone
Power Source: Single Frequency
Press F1 for more help.
Select: Enter the starting point

Fig. 5.76 .

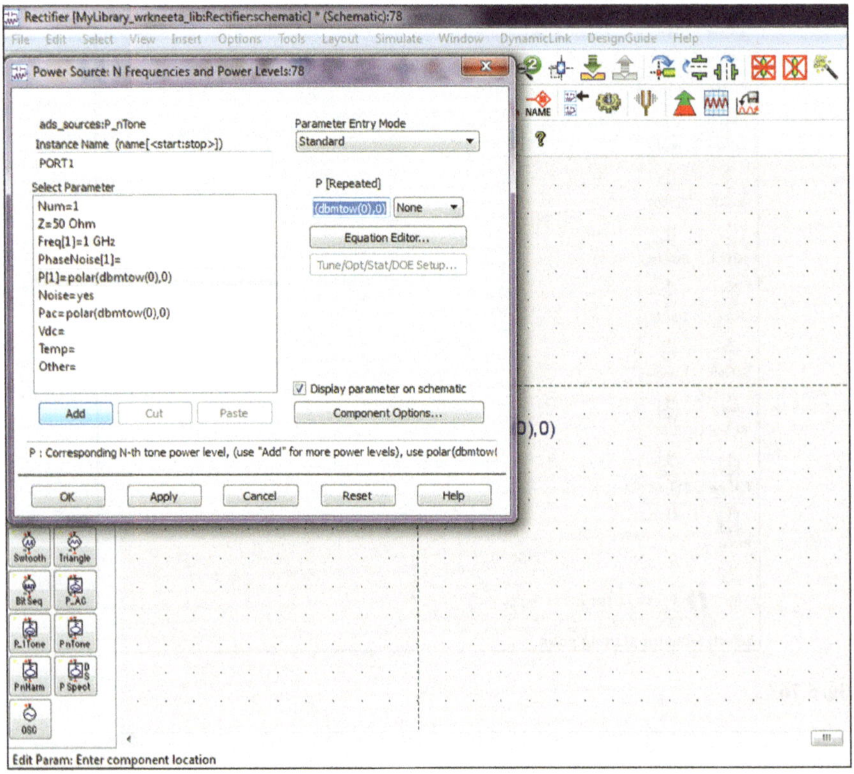
Rectifier [MyLibrary_wrkneeta_lib:Rectifier:schematic] * (Schematic):78
File Edit Select View Insert Options Tools Layout Simulate Window DynamicLink DesignGuide Help
Power Source: N Frequencies and Power Levels:78
ads_sources:P_nTone
Instance Name (name[<start:stop>])
PORT1
Select Parameter
Num=1
Z=50 Ohm
Freq[1]=1 GHz
PhaseNoise[1]=
P[1]=polar(dbmtow(0),0)
Noise=yes
Pac=polar(dbmtow(0),0)
Vdc=
Temp=
Other=
Add
Cut
Paste
Parameter Entry Mode
Standard
P [Repeated]
(dbmtow(0),0)
None
Equation Editor...
Tune/Opt/Stat/DOE Setup...
Display parameter on schematic
Component Options...
P : Corresponding N-th tone power level, (use "Add" for more power levels), use polar(dbmtow(
OK
Apply
Cancel
Reset
Help
Swtooth
Triangle
Bit Seq
P_AC
P_1Tone
PnTone
PnHarm
P Spect
OSC
Edit Param: Enter component location

Fig. 5.77 .

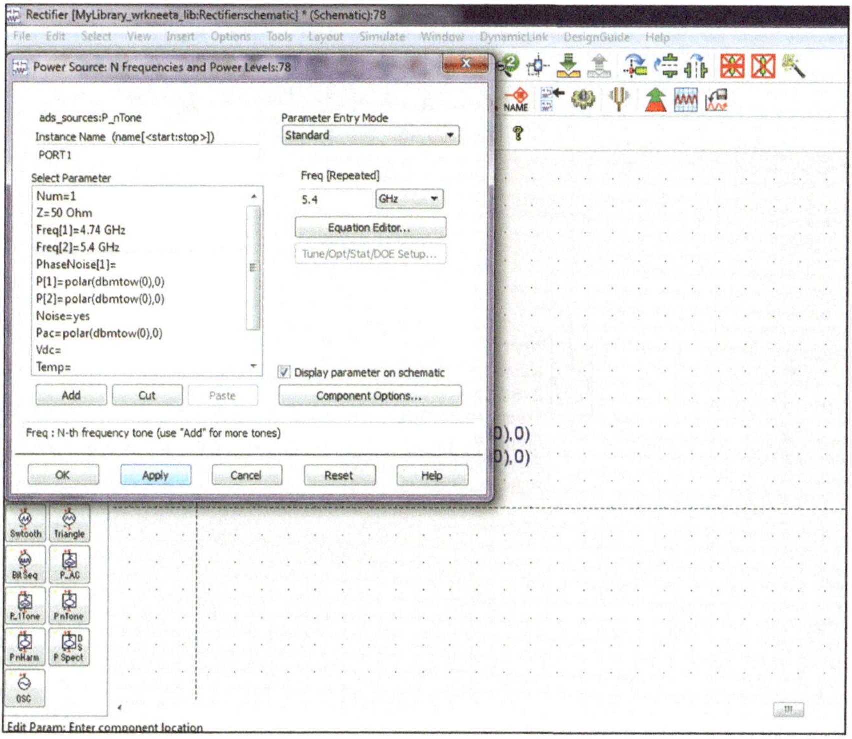
Rectifier [MyLibrary_wrkneeta_lib:Rectifier:schematic] * (Schematic):78
File Edit Select View Insert Options Tools Layout Simulate Window DynamicLink DesignGuide Help
Power Source: N Frequencies and Power Levels:78
ads_sources:P_nTone
Instance Name (name[<start:stop>])
PORT1
Parameter Entry Mode
Standard
Select Parameter
Num=1
Z=50 Ohm
Freq[1]=4.74 GHz
Freq[2]=5.4 GHz
PhaseNoise[1]=
P[1]=polar(dbmtow(0),0)
P[2]=polar(dbmtow(0),0)
Noise=yes
Pac=polar(dbmtow(0),0)
Vdc=
Temp=
Freq [Repeated]
5.4
GHz
Equation Editor...
Tune/Opt/Stat/DOE Setup...
Display parameter on schematic
Add
Cut
Paste
Component Options...
Freq : N-th frequency tone (use "Add" for more tones)
OK
Apply
Cancel
Reset
Help
NAME
Swtooth
Triangle
BitSeq
P_AC
P_1Tone
PnTone
PnHarm
P Spect
OSC
Edit Param: Enter component location

Fig. 5.78 .

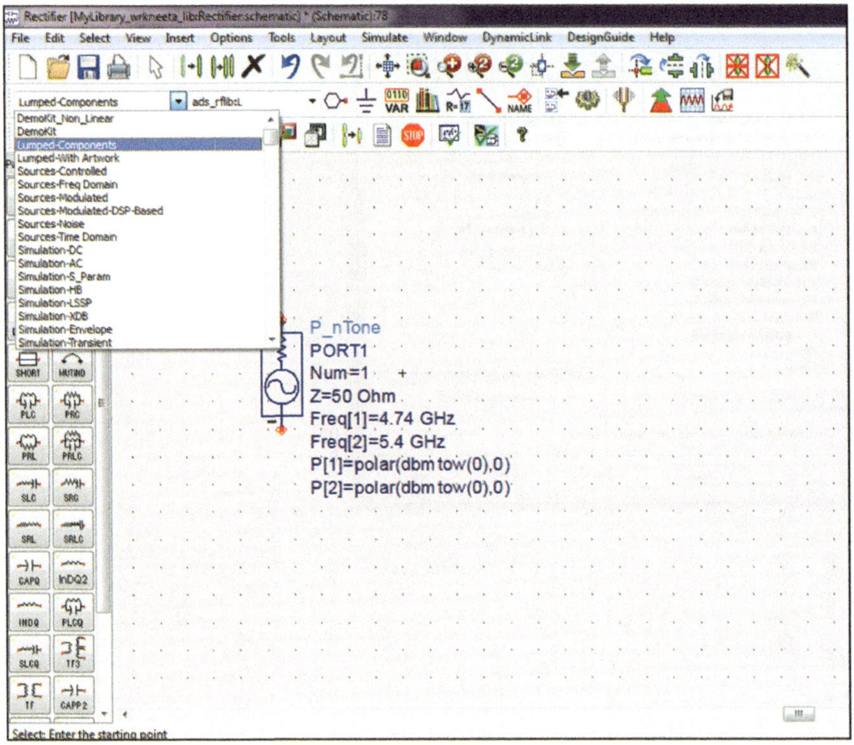
Rectifier [MyLibrary_wrkneeta_lib:Rectifier:schematic] * (Schematic):78
File Edit Select View Insert Options Tools Layout Simulate Window DynamicLink DesignGuide Help
Lumped-Components
ads_rflib:L
DemoKit_Non_Linear
DemoKit
Lumped-Components
Lumped-With Artwork
Sources-Controlled
Sources-Freq Domain
Sources-Modulated
Sources-Modulated-DSP-Based
Sources-Noise
Sources-Time Domain
Simulation-DC
Simulation-AC
Simulation-S_Param
Simulation-HB
Simulation-LSSP
Simulation-XDB
Simulation-Envelope
Simulation-Transient
P_nTone
PORT1
Num=1
Z=50 Ohm
Freq[1]=4.74 GHz
Freq[2]=5.4 GHz
P[1]=polar(dbmtow(0),0)
P[2]=polar(dbmtow(0),0)
Select: Enter the starting point

Fig. 5.79 .

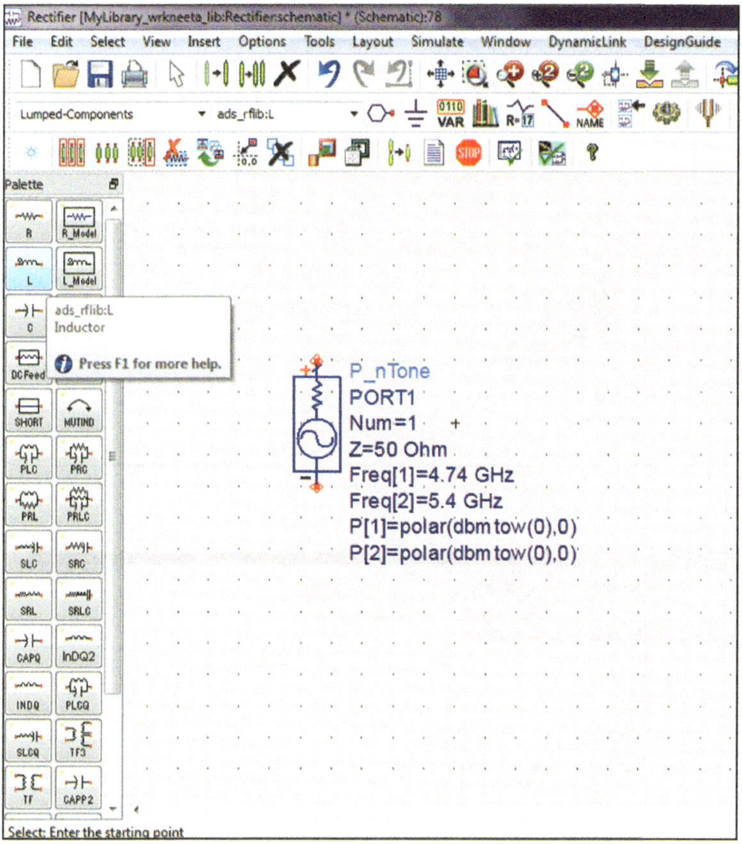
Rectifier [MyLibrary_wrkneeta_lib:Rectifier:schematic] * (Schematic):78
File Edit Select View Insert Options Tools Layout Simulate Window DynamicLink DesignGuide
Lumped-Components
ads_rflib:L
VAR
NAME
Palette
R
R_Model
L
L_Model
C
DCFeed
ads_rflib:L
Inductor
Press F1 for more help.
SHORT
MUTIND
PLC
PRC
PRL
PRLC
SLC
SRC
SRL
SRLC
CAPQ
InDQ2
INDQ
PLCQ
SLCQ
TF3
TF
CAPP2
P_nTone
PORT1
Num=1
Z=50 Ohm
Freq[1]=4.74 GHz
Freq[2]=5.4 GHz
P[1]=polar(dbmtow(0),0)
P[2]=polar(dbmtow(0),0)
Select: Enter the starting point

Fig. 5.80 .

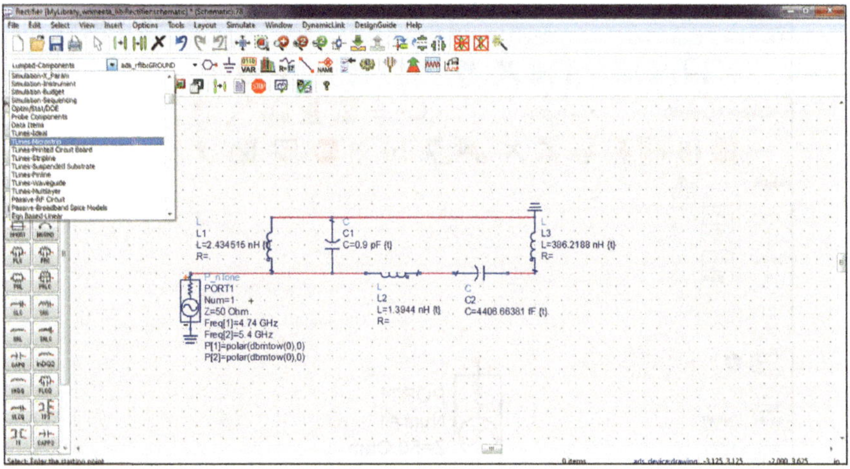

Fig. 5.81 .

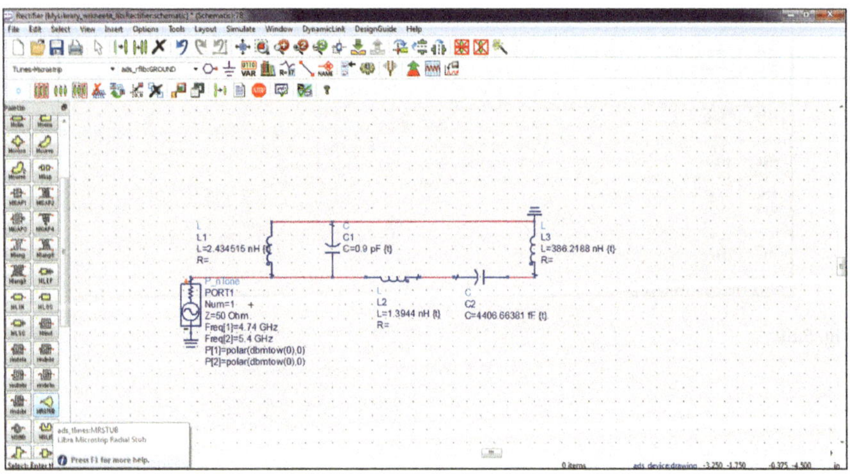

Fig. 5.82 .

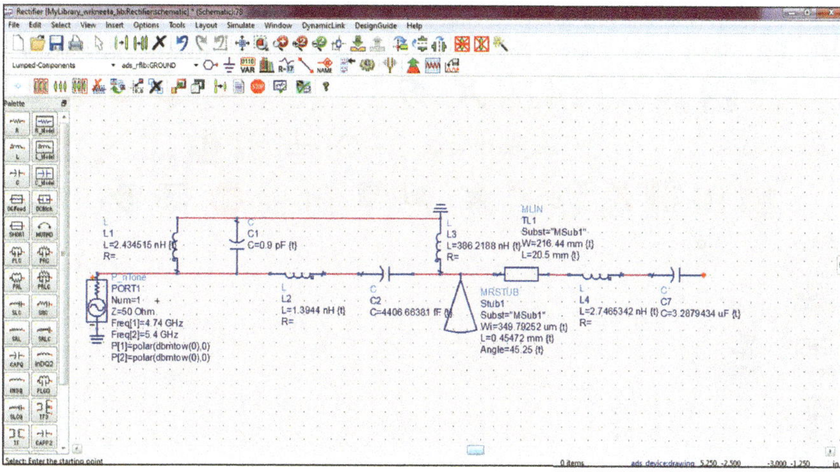

Fig. 5.83 .

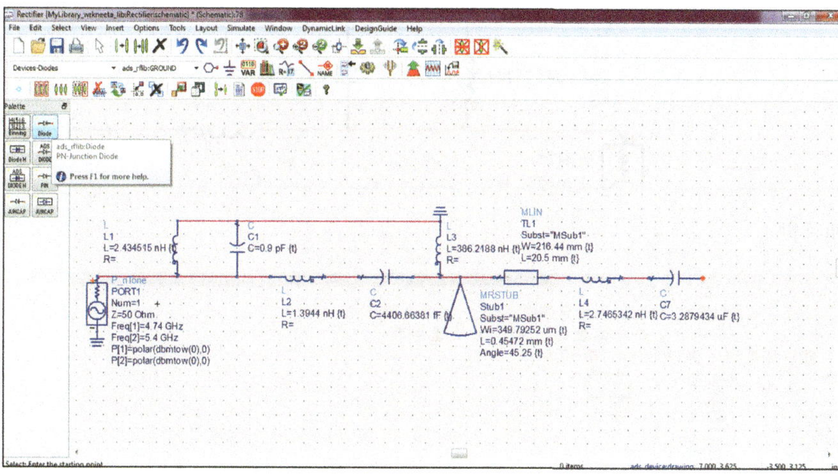

Fig. 5.84 .

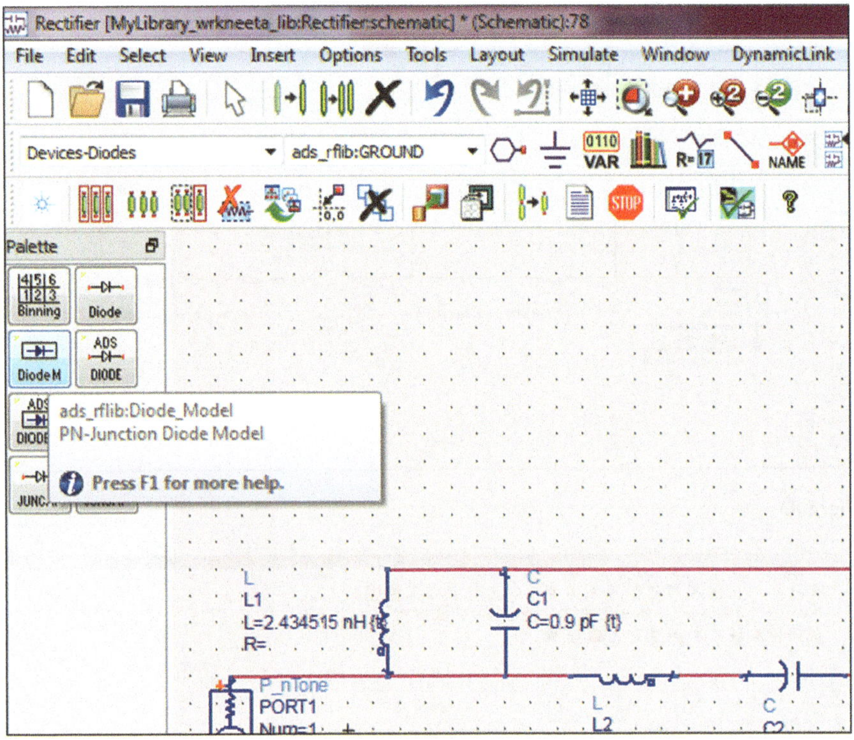

Fig. 5.85 .

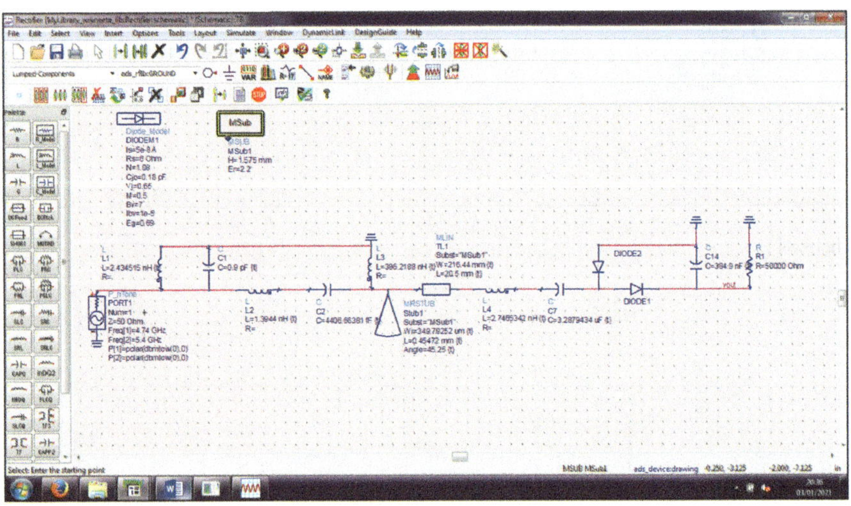

Fig. 5.86 .

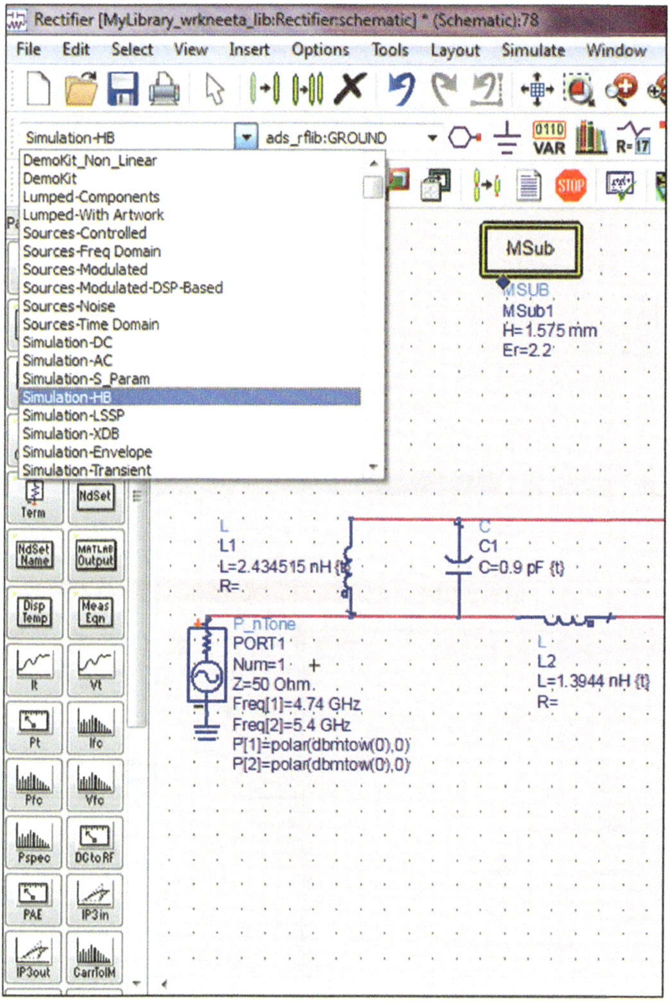
Rectifier [MyLibrary_wrkneeta_lib:Rectifier:schematic] * (Schematic):78
File Edit Select View Insert Options Tools Layout Simulate Window
Simulation-HB
ads_rflib:GROUND
DemoKit_Non_Linear
DemoKit
Lumped-Components
Lumped-With Artwork
Sources-Controlled
Sources-Freq Domain
Sources-Modulated
Sources-Modulated-DSP-Based
Sources-Noise
Sources-Time Domain
Simulation-DC
Simulation-AC
Simulation-S_Param
Simulation-HB
Simulation-LSSP
Simulation-XDB
Simulation-Envelope
Simulation-Transient
MSub
MSUB
MSub1
H=1.575 mm
Er=2.2
L
L1
L=2.434515 nH {t}
R=
C
C1
C=0.9 pF {t}
P_nTone
PORT1
Num=1
Z=50 Ohm
Freq[1]=4.74 GHz
Freq[2]=5.4 GHz
P[1]=polar(dbmtow(0),0)
P[2]=polar(dbmtow(0),0)
L
L2
L=1.3944 nH {t}
R=
Term
NdSet
NdSet Name
MATLAB Output
Disp Temp
Meas Eqn
It
Vt
Pt
Ifc
Pfc
Vfc
Pspec
DCtoRF
PAE
IP3in
IP3out
CarrToIM

Fig. 5.87 .

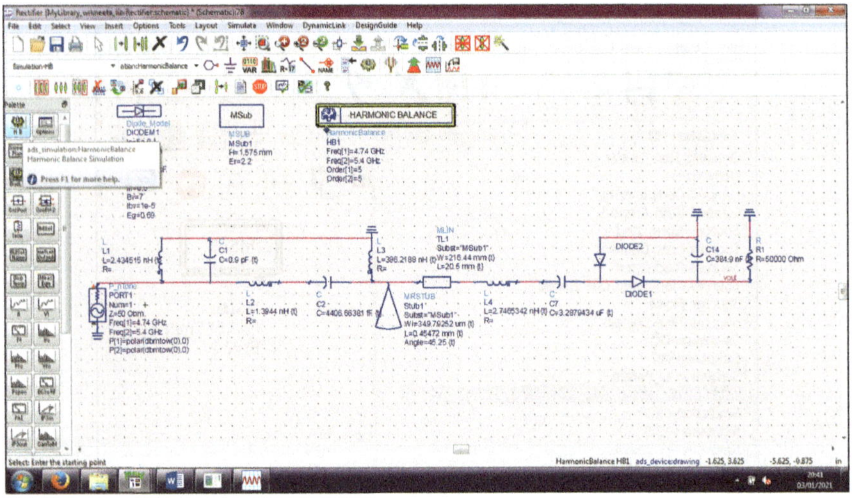

Fig. 5.88 .

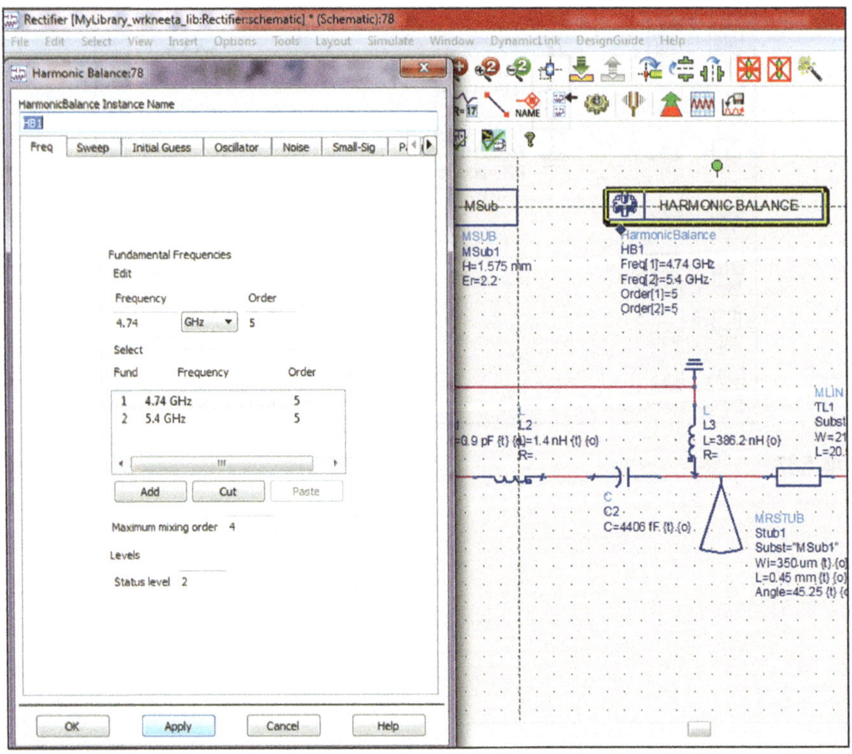

Fig. 5.89 .

Fig. 5.90 .

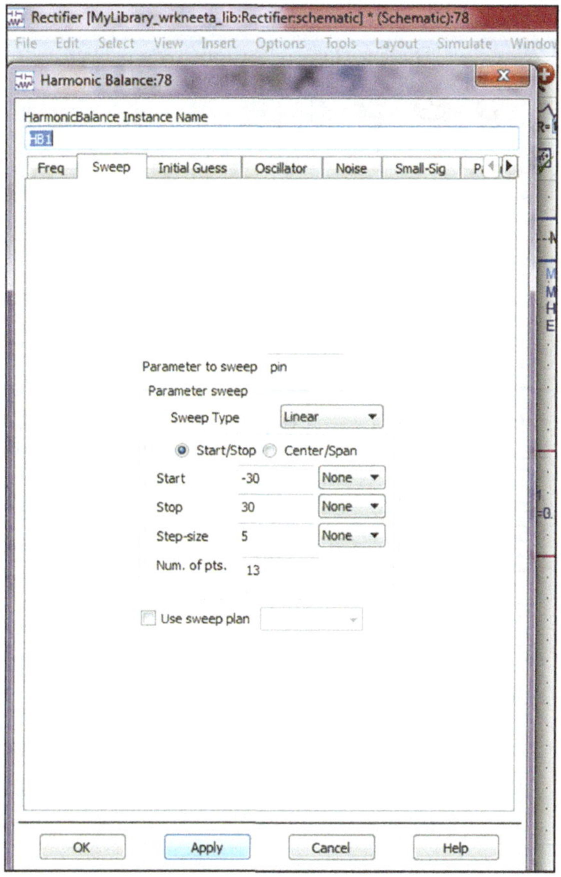
Rectifier [MyLibrary_wrkneeta_lib:Rectifier:schematic] * (Schematic):78
File Edit Select View Insert Options Tools Layout Simulate
Harmonic Balance:78
HarmonicBalance Instance Name
HB1
Freq
Sweep
Initial Guess
Oscillator
Noise
Small-Sig
Parameter to sweep pin
Parameter sweep
Sweep Type Linear
Start/Stop Center/Span
Start -30 None
Stop 30 None
Step-size 5 None
Num. of pts. 13
Use sweep plan
OK
Apply
Cancel
Help

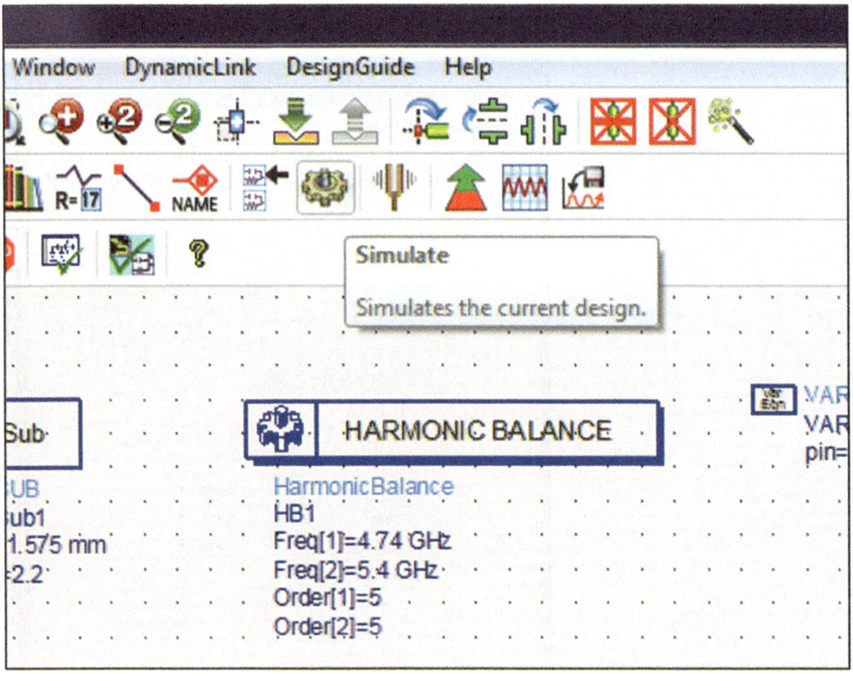

Fig. 5.91 .

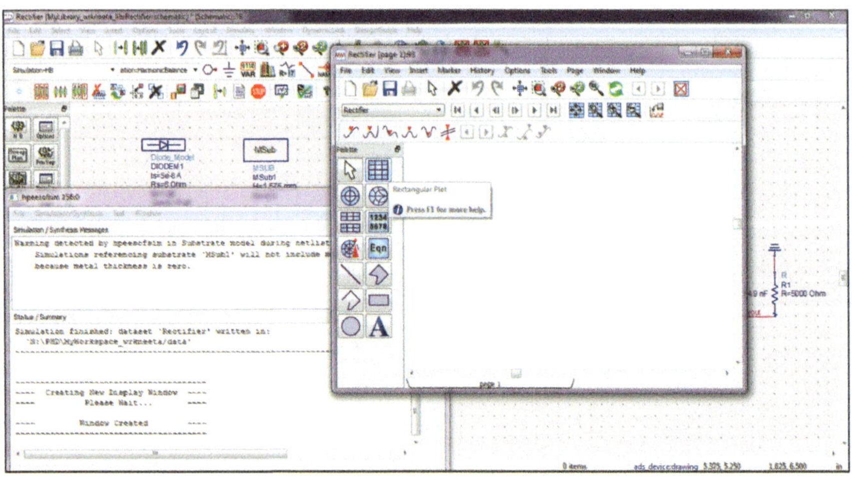

Fig. 5.92 .

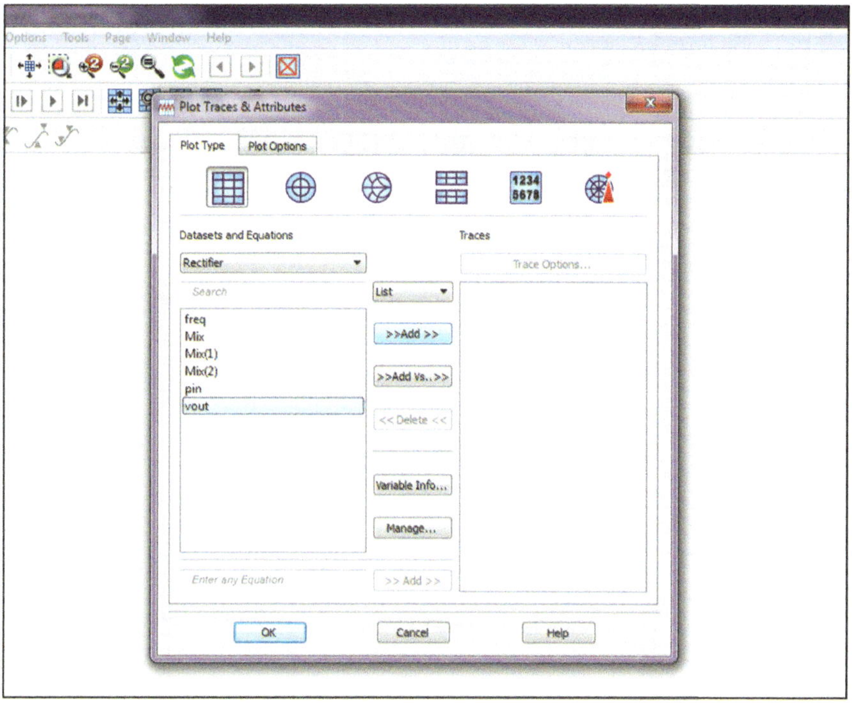
Options Tools Page Window Help
Plot Traces & Attributes
Plot Type
Plot Options
Datasets and Equations
Traces
Rectifier
Trace Options...
Search
List
freq
Mix
Mix(1)
Mix(2)
pin
vout
>>Add >>
>>Add Vs..>>
<< Delete <<
Variable Info...
Manage...
Enter any Equation
>> Add >>
OK
Cancel
Help

Fig. 5.93 .

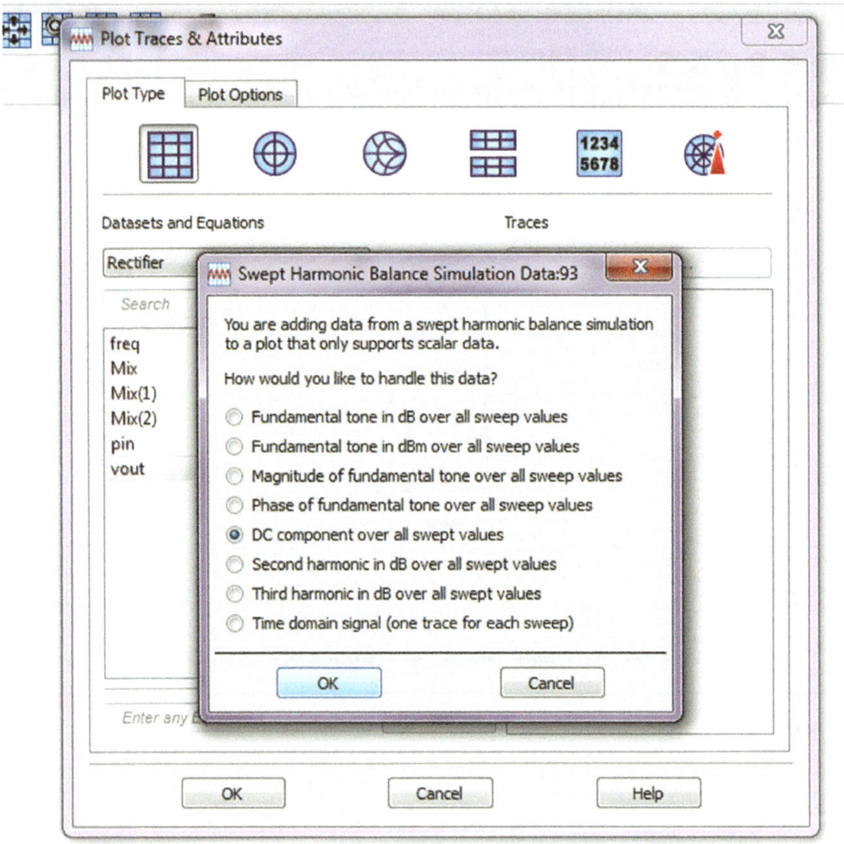
Plot Traces & Attributes
Plot Type
Plot Options
1234
5678
Datasets and Equations
Traces
Rectifier
Search
freq
Mix
Mix(1)
Mix(2)
pin
vout
Swept Harmonic Balance Simulation Data:93
You are adding data from a swept harmonic balance simulation to a plot that only supports scalar data.
How would you like to handle this data?
Fundamental tone in dB over all sweep values
Fundamental tone in dBm over all sweep values
Magnitude of fundamental tone over all sweep values
Phase of fundamental tone over all sweep values
DC component over all swept values
Second harmonic in dB over all swept values
Third harmonic in dB over all swept values
Time domain signal (one trace for each sweep)
OK
Cancel
OK
Cancel
Help

Fig. 5.94 .

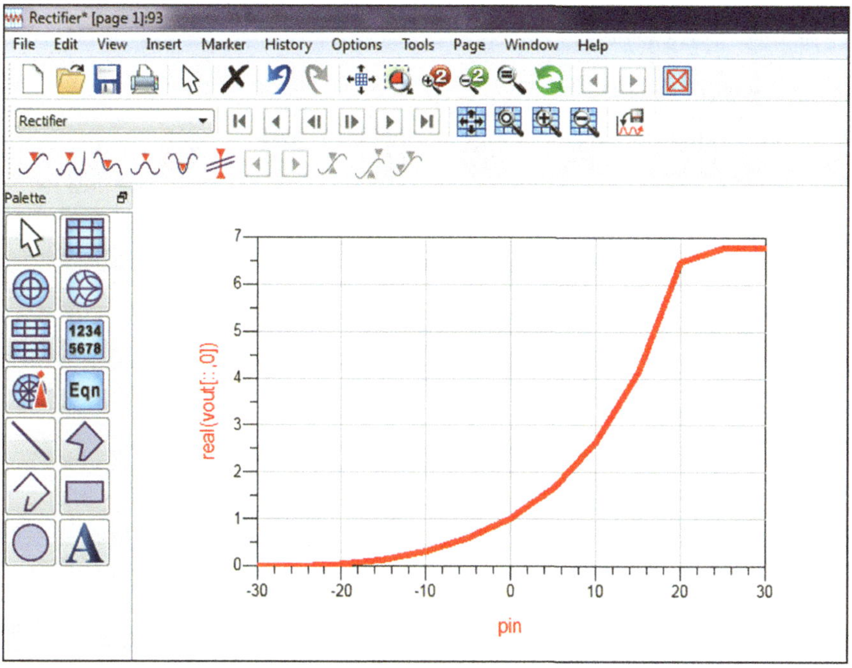

Fig. 5.95 .

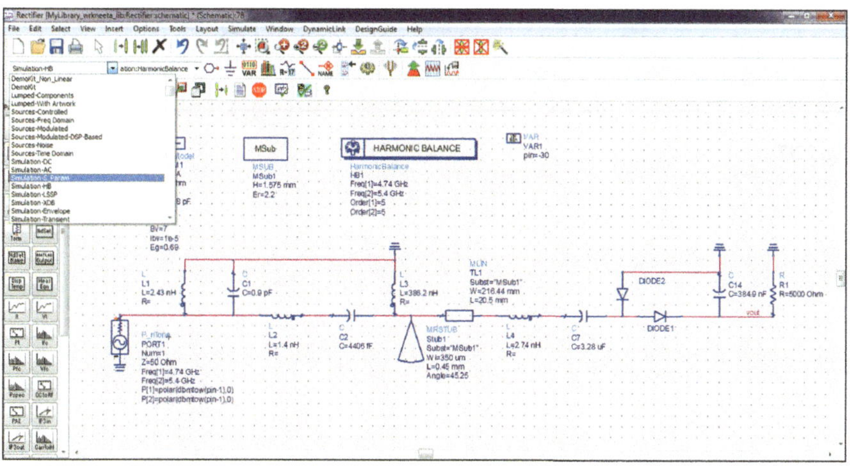

Fig. 5.96 .

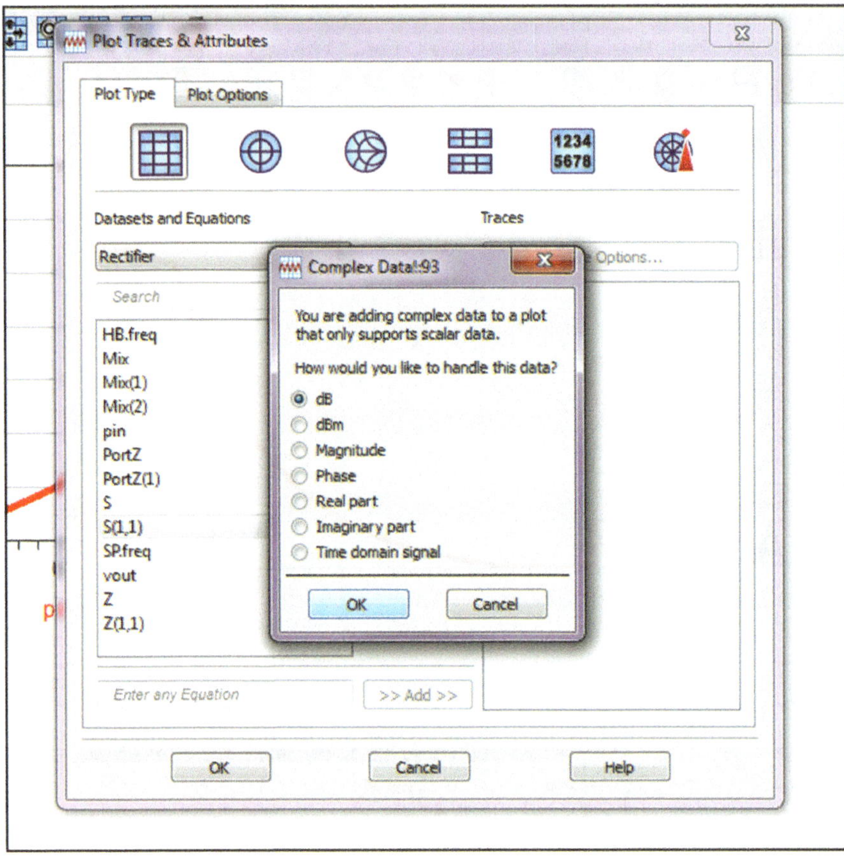
Plot Traces & Attributes
Plot Type
Plot Options
1234
5678
Datasets and Equations
Traces
Rectifier
Options...
Search
HB.freq
Mix
Mix(1)
Mix(2)
pin
PortZ
PortZ(1)
S
S(1,1)
SP.freq
vout
Z
Z(1,1)
Complex Data!:93
You are adding complex data to a plot that only supports scalar data.
How would you like to handle this data?
dB
dBm
Magnitude
Phase
Real part
Imaginary part
Time domain signal
OK
Cancel
Enter any Equation
>> Add >>
OK
Cancel
Help

Fig. 5.97 .

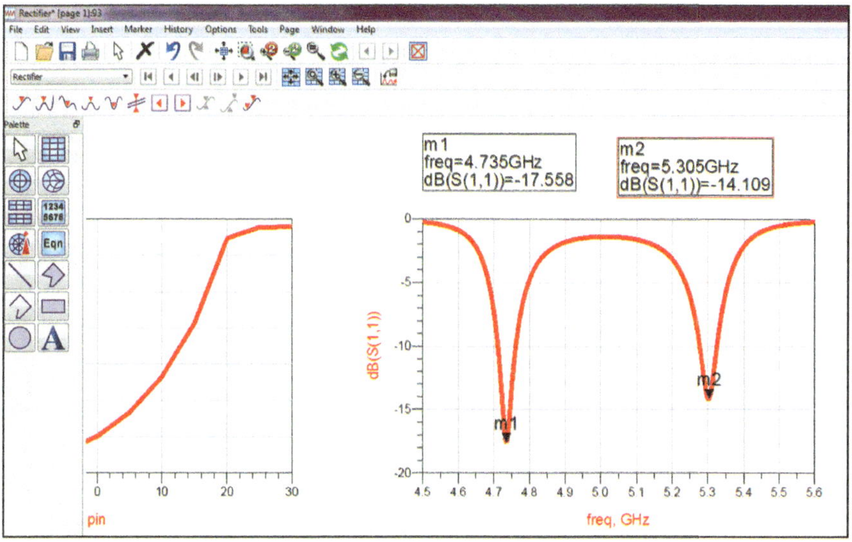

Fig. 5.98 .

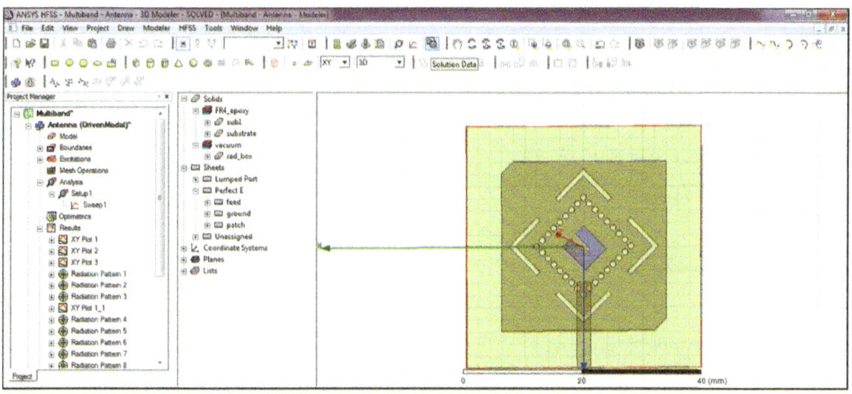

Fig. 5.99 .

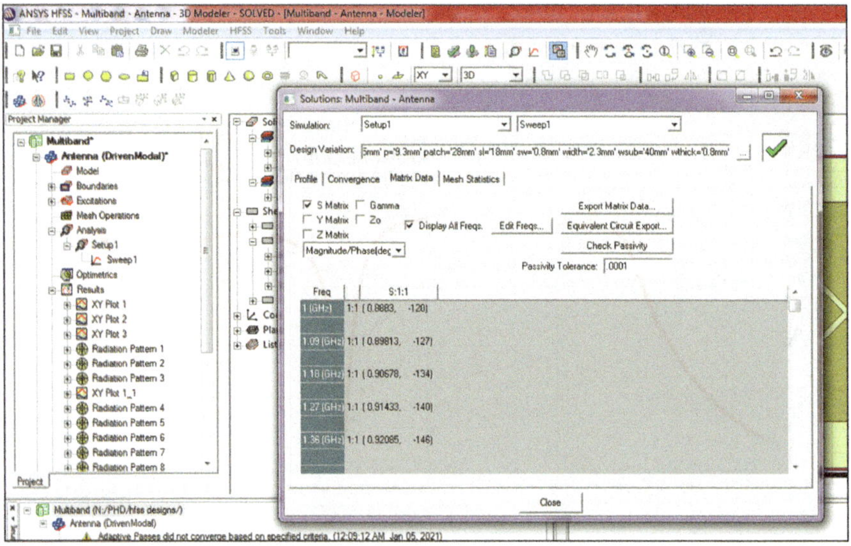

Fig. 5.100 .

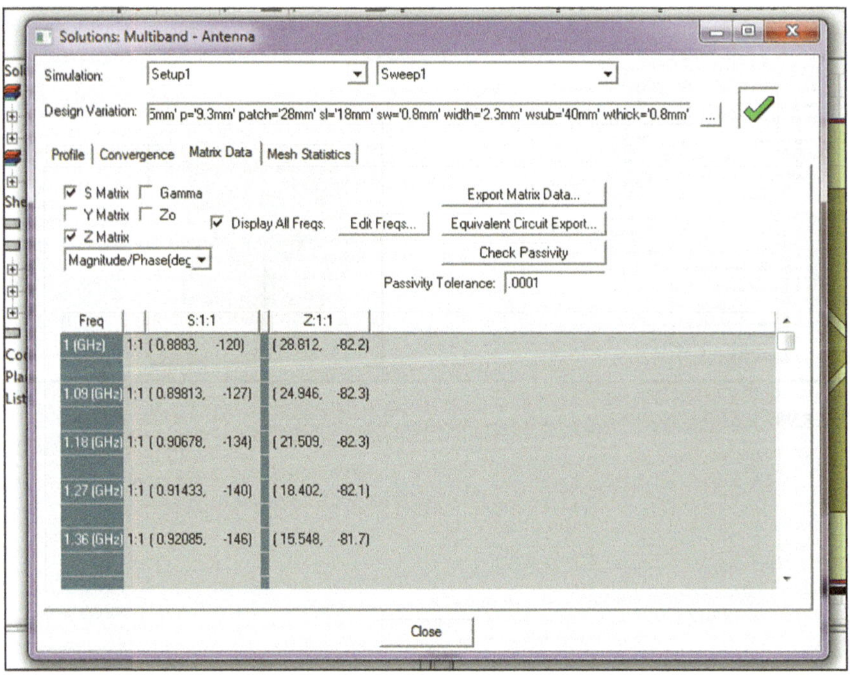

Fig. 5.101 .

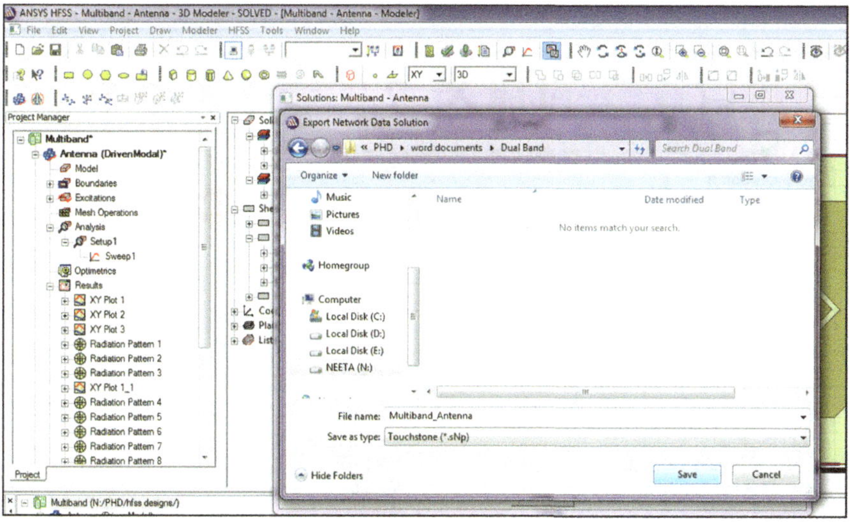

Fig. 5.102 .

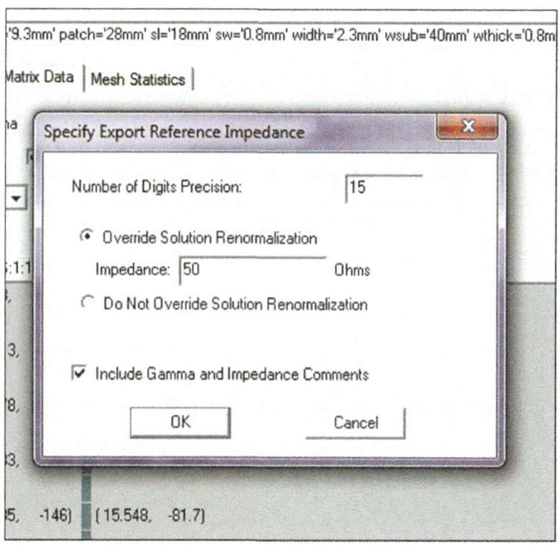

Fig. 5.103 .

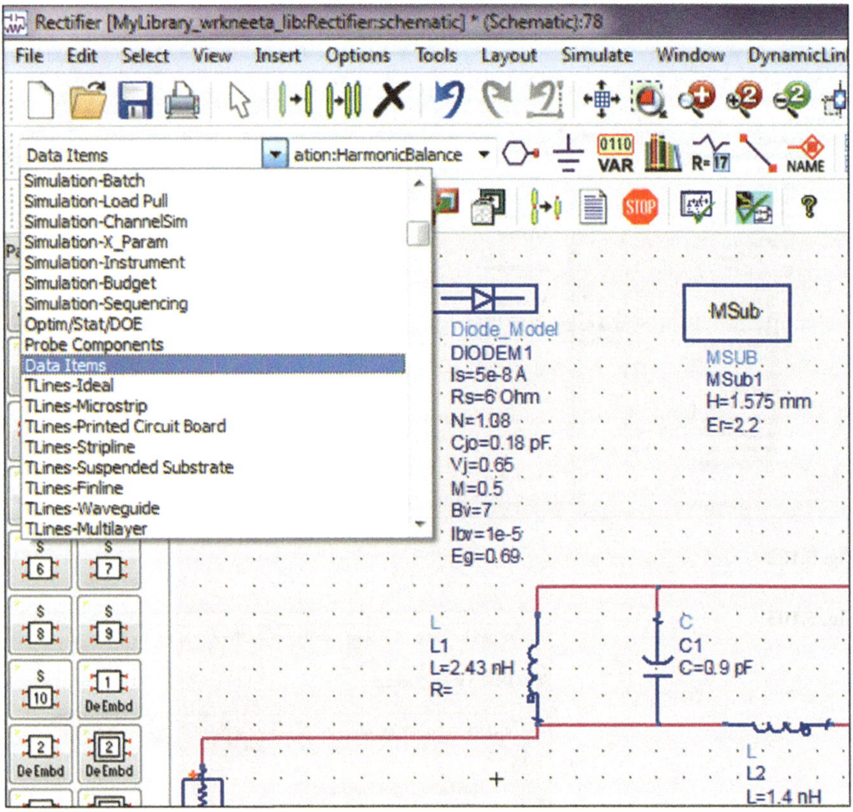
Rectifier [MyLibrary_wrkneeta_lib:Rectifier:schematic] * (Schematic):78
File
Edit
Select
View
Insert
Options
Tools
Layout
Simulate
Window
DynamicLin
Data Items
ation:HarmonicBalance
VAR
NAME
Simulation-Batch
Simulation-Load Pull
Simulation-ChannelSim
Simulation-X_Param
Simulation-Instrument
Simulation-Budget
Simulation-Sequencing
Optim/Stat/DOE
Probe Components
Data Items
TLines-Ideal
TLines-Microstrip
TLines-Printed Circuit Board
TLines-Stripline
TLines-Suspended Substrate
TLines-Finline
TLines-Waveguide
TLines-Multilayer
Diode_Model
DIODEM1
Is=5e-8 A
Rs=6 Ohm
N=1.08
Cjo=0.18 pF
Vj=0.65
M=0.5
Bv=7
Ibv=1e-5
Eg=0.69
MSub
MSUB
MSub1
H=1.575 mm
Er=2.2
L
L1
L=2.43 nH
R=
C
C1
C=0.9 pF
L
L2
L=1.4 nH
DeEmbd

Fig. 5.104 .

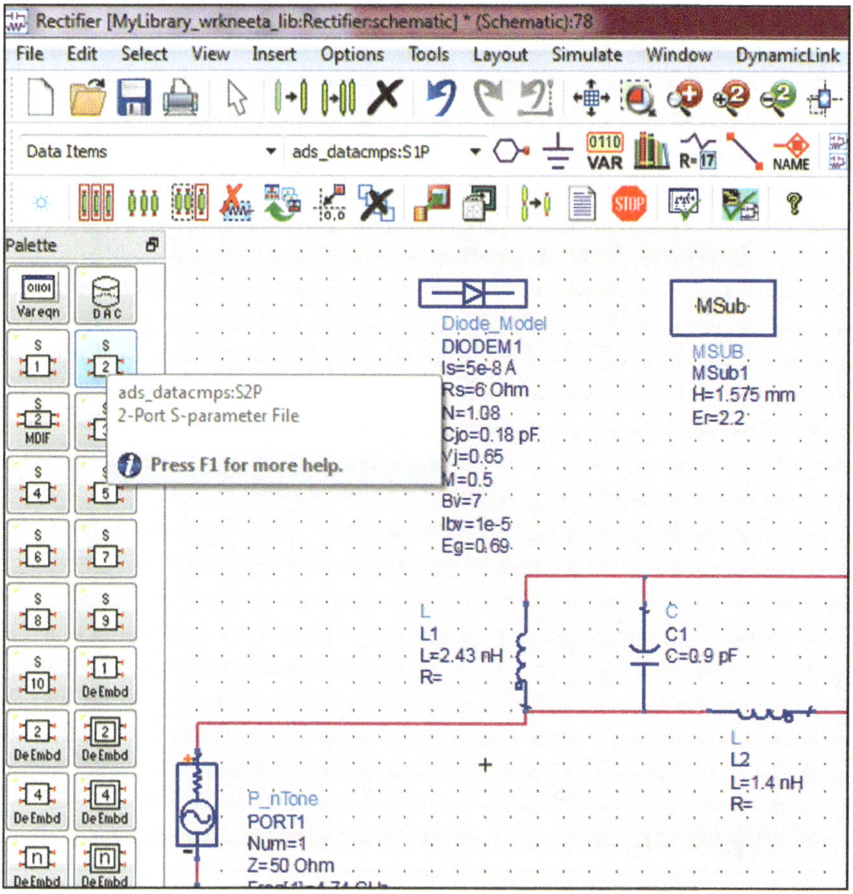
Rectifier [MyLibrary_wrkneeta_lib:Rectifier:schematic] * (Schematic):78
File Edit Select View Insert Options Tools Layout Simulate Window DynamicLink
Data Items
ads_datacmps:S1P
VAR
NAME
Palette
Var eqn
DAC
MDIF
De Embd
ads_datacmps:S2P
2-Port S-parameter File
Press F1 for more help.
Diode_Model
DIODEM1
Is=5e-8 A
Rs=6 Ohm
N=1.08
Cjo=0.18 pF
Vj=0.65
M=0.5
Bv=7
Ibv=1e-5
Eg=0.69
MSub
MSUB
MSub1
H=1.575 mm
Er=2.2
L
L1
L=2.43 nH
R=
C
C1
C=0.9 pF
L
L2
L=1.4 nH
R=
P_nTone
PORT1
Num=1
Z=50 Ohm

Fig. 5.105 .

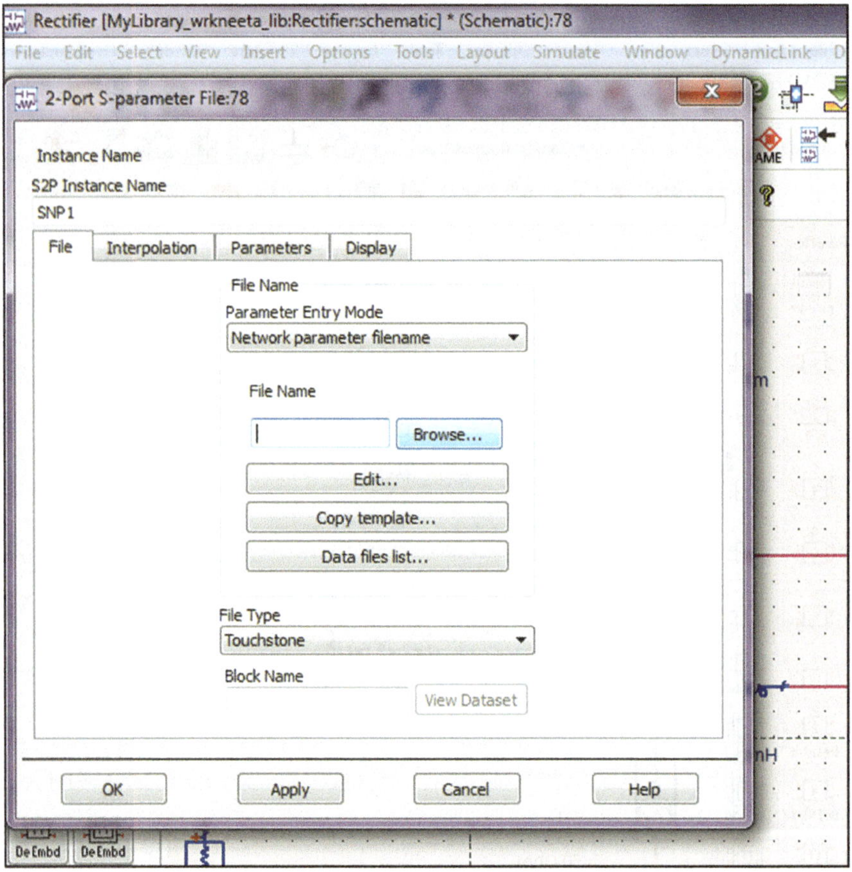
Rectifier [MyLibrary_wrkneeta_lib:Rectifier:schematic] * (Schematic):78
File Edit Select View Insert Options Tools Layout Simulate Window DynamicLink
2-Port S-parameter File:78
Instance Name
S2P Instance Name
SNP1
File
Interpolation
Parameters
Display
File Name
Parameter Entry Mode
Network parameter filename
File Name
Browse...
Edit...
Copy template...
Data files list...
File Type
Touchstone
Block Name
View Dataset
OK
Apply
Cancel
Help
DeEmbd
DeEmbd

Fig. 5.106 .

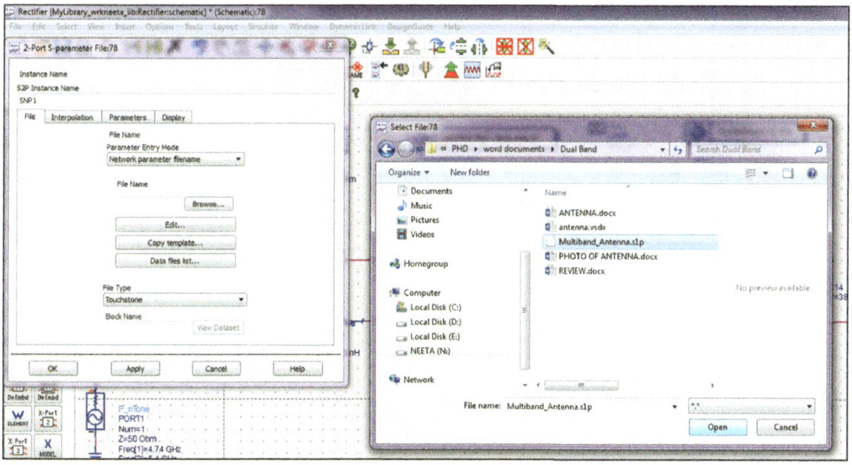

Fig. 5.107 .

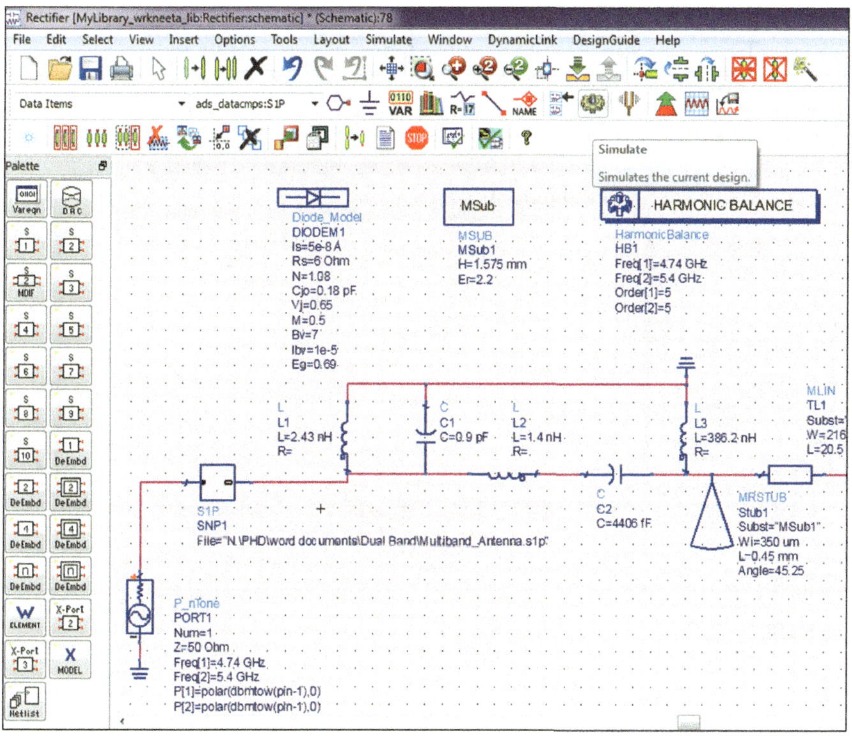

Fig. 5.108 .

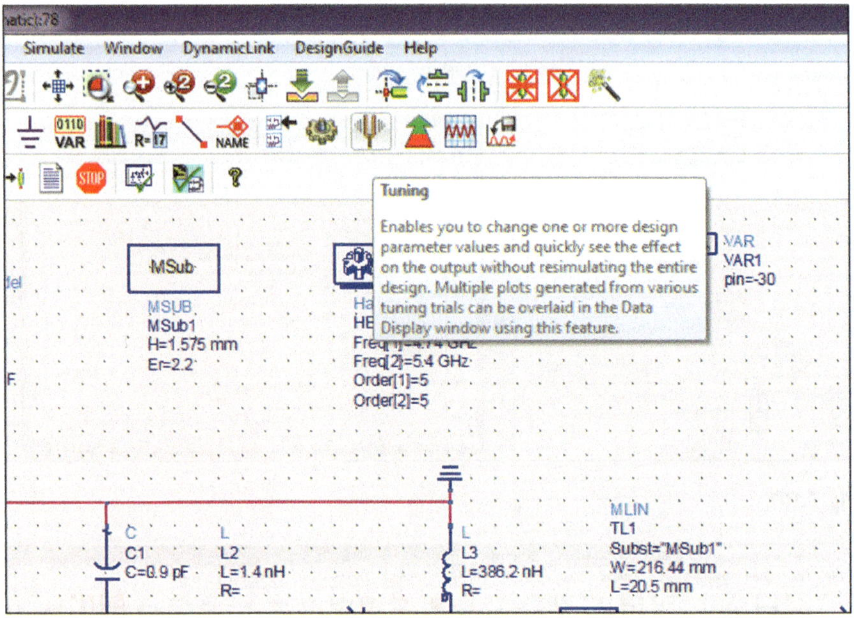

Fig. 5.109 .

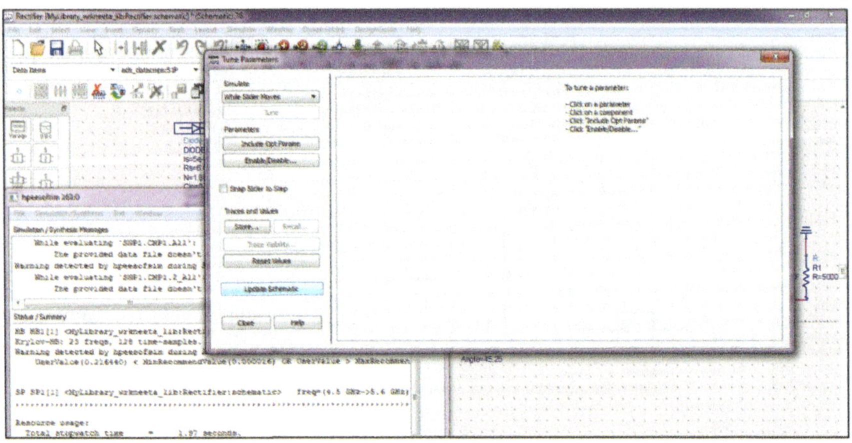

Fig. 5.110 .

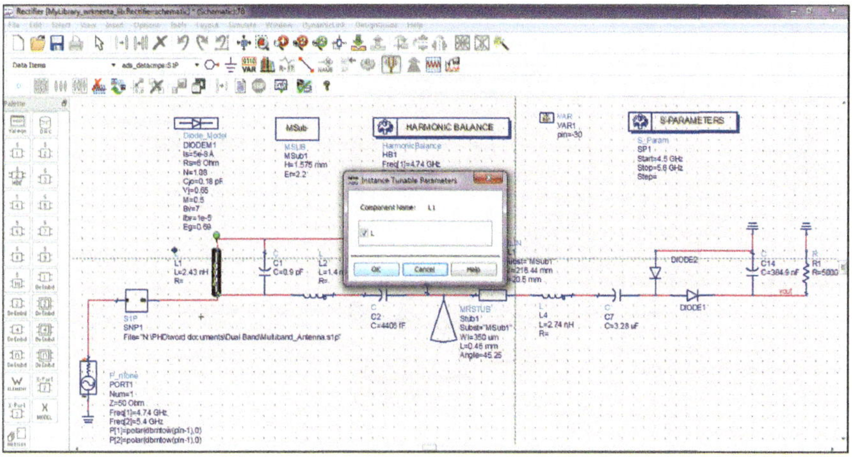

Fig. 5.111 .

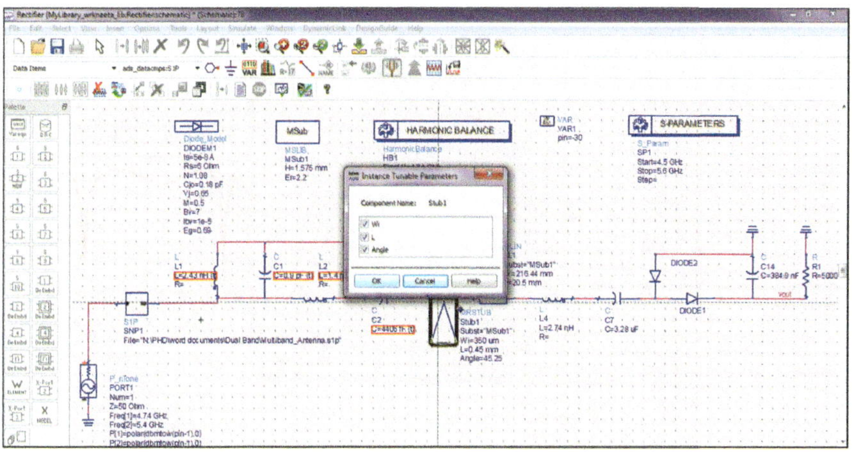

Fig. 5.112 .

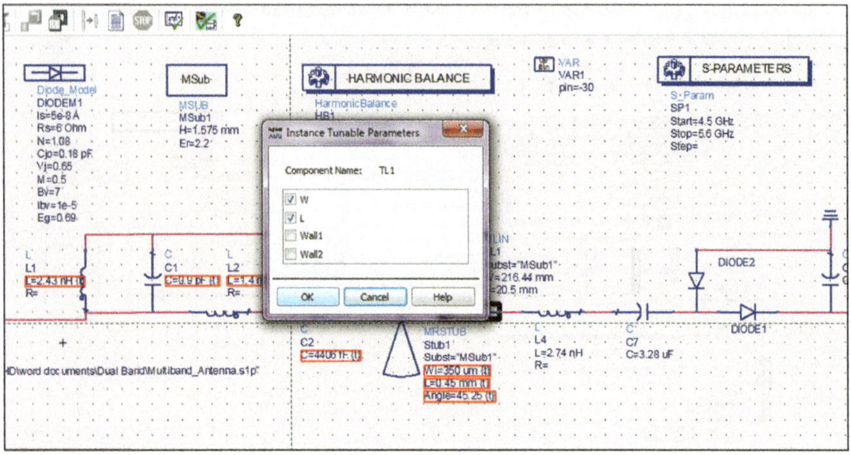

Fig. 5.113 .

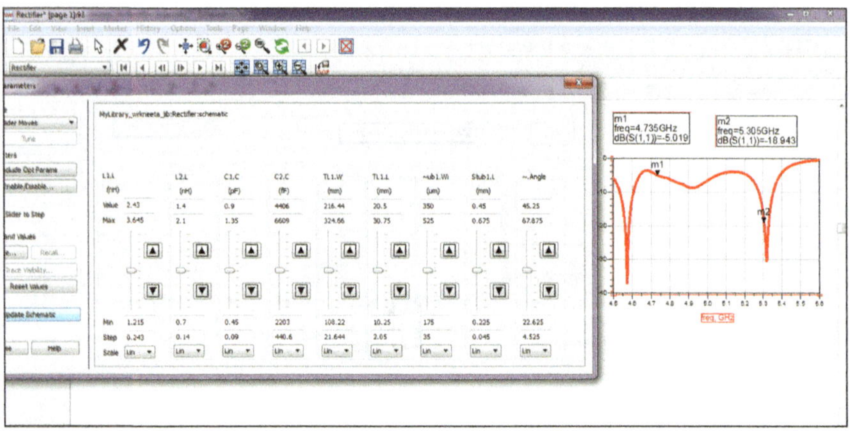

Fig. 5.114 .

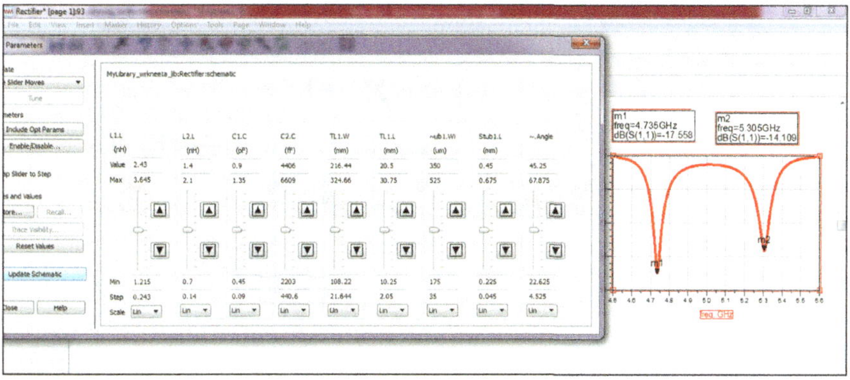

Fig. 5.115 .

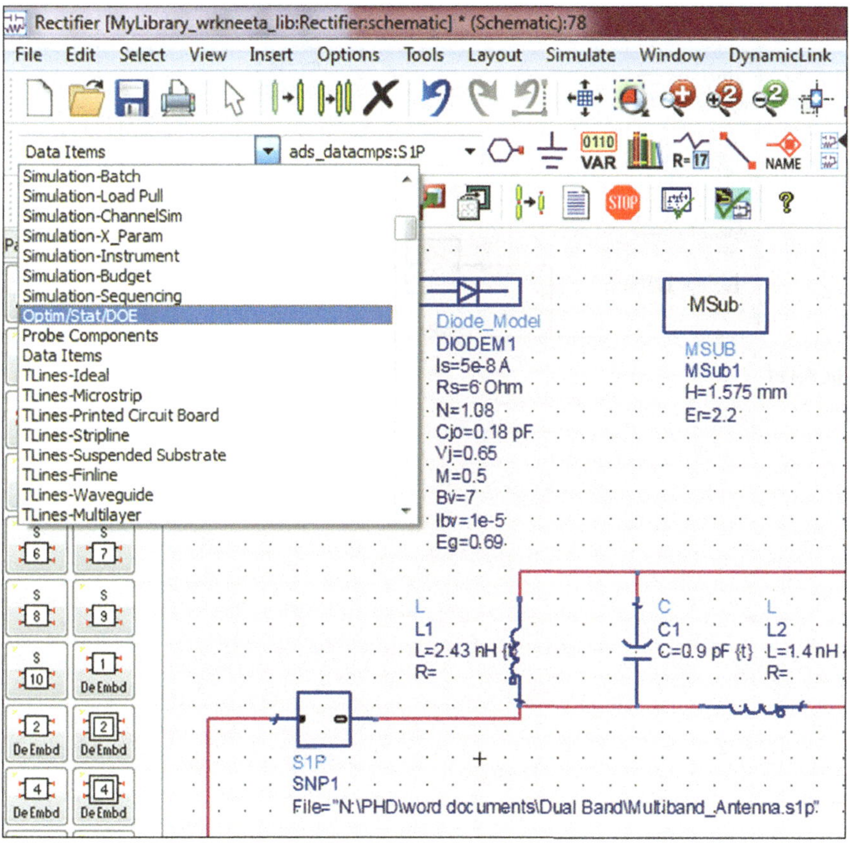

Fig. 5.116 .

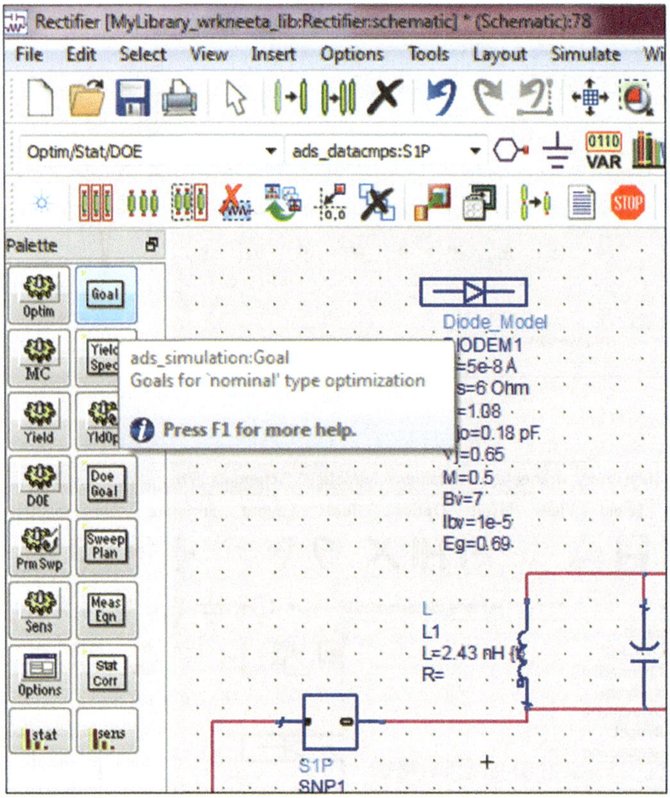
Rectifier [MyLibrary_wrkneeta_lib:Rectifier:schematic] * (Schematic):78
File Edit Select View Insert Options Tools Layout Simulate
Optim/Stat/DOE
ads_datacmps:S1P
VAR
Palette
Optim
Goal
MC
Yield
DOE
Doe Goal
Prm Swp
Sweep Plan
Sens
Meas Eqn
Options
Stat Corr
stat
sens
ads_simulation:Goal
Goals for `nominal' type optimization
Press F1 for more help.
Diode_Model
M=0.5
Bv=7
Ibv=1e-5
Eg=0.69
L
L1
L=2.43 nH
R=
S1P
SNP1

Fig. 5.117 .

Fig. 5.118 .

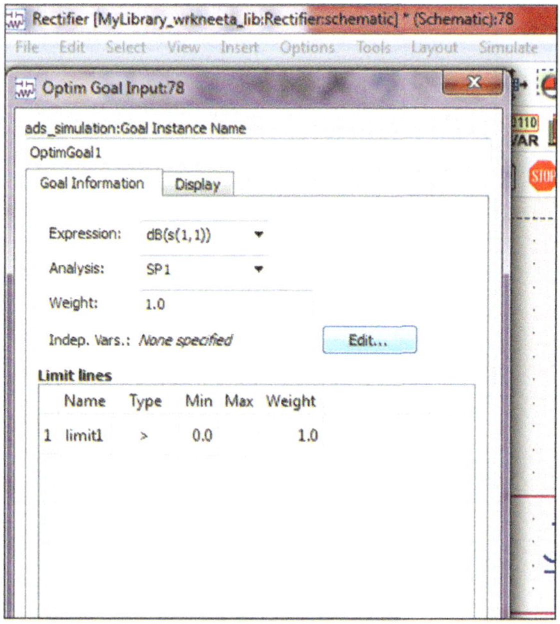
Rectifier [MyLibrary_wrkneeta_lib:Rectifier:schematic] * (Schematic):78
File Edit Select View Insert Options Tools Layout Simulate
Optim Goal Input:78
ads_simulation:Goal Instance Name
OptimGoal1
Goal Information
Display
Expression: dB(s(1,1))
Analysis: SP1
Weight: 1.0
Indep. Vars.: None specified
Edit...
Limit lines
Name Type Min Max Weight
1 limit1 > 0.0 1.0

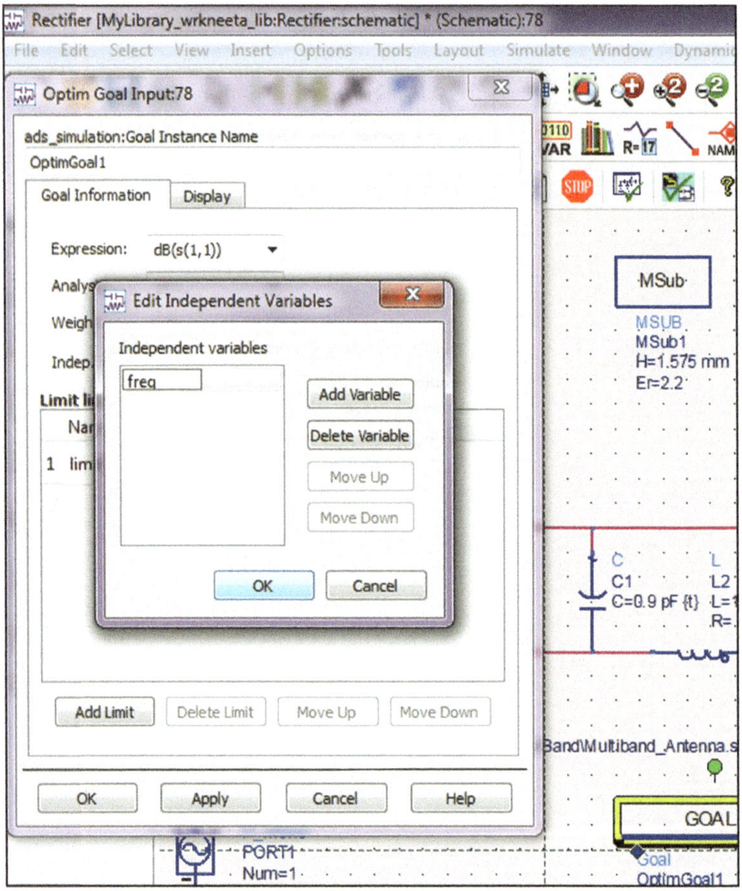
Rectifier [MyLibrary_wrkneeta_lib:Rectifier:schematic] * (Schematic):78
File Edit Select View Insert Options Tools Layout Simulate Window Dynamic
Optim Goal Input:78
ads_simulation:Goal Instance Name
OptimGoal1
Goal Information
Display
Expression:
dB(s(1,1))
Edit Independent Variables
Independent variables
freq
Add Variable
Delete Variable
Move Up
Move Down
OK
Cancel
Add Limit
Delete Limit
Move Up
Move Down
OK
Apply
Cancel
Help
MSub
MSUB
MSub1
H=1.575 mm
Er=2.2
C
C1
C=0.9 pF {t}
L
L2
Band\Multiband_Antenna.s
GOAL
PORT1
Num=1
Goal
OptimGoal1

Fig. 5.119 .

Fig. 5.120 .

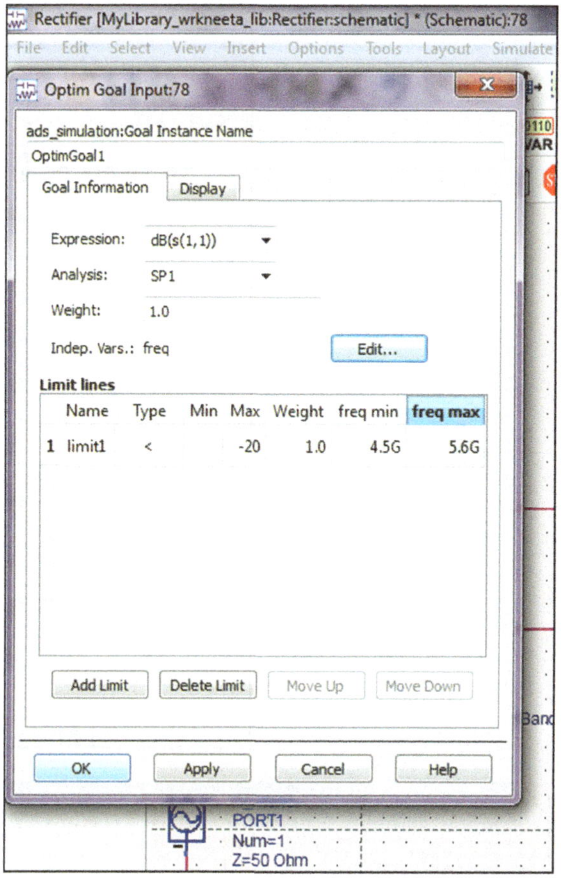
Rectifier [MyLibrary_wrkneeta_lib:Rectifier:schematic] * (Schematic):78
File Edit Select View Insert Options Tools Layout Simulate
Optim Goal Input:78
ads_simulation:Goal Instance Name
OptimGoal1
Goal Information
Display
Expression: dB(s(1,1))
Analysis: SP1
Weight: 1.0
Indep. Vars.: freq
Edit...
Limit lines
Name Type Min Max Weight freq min freq max
1 limit1 < -20 1.0 4.5G 5.6G
Add Limit
Delete Limit
Move Up
Move Down
OK
Apply
Cancel
Help
PORT1
Num=1
Z=50 Ohm

Fig. 5.121 .

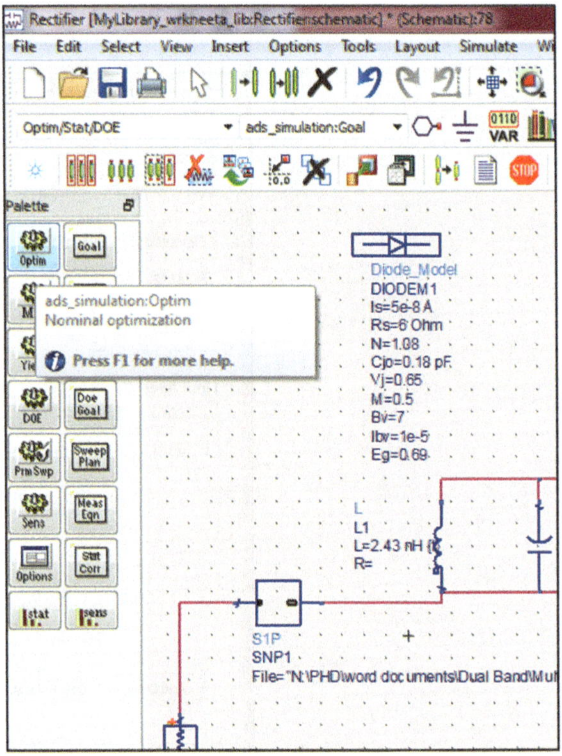

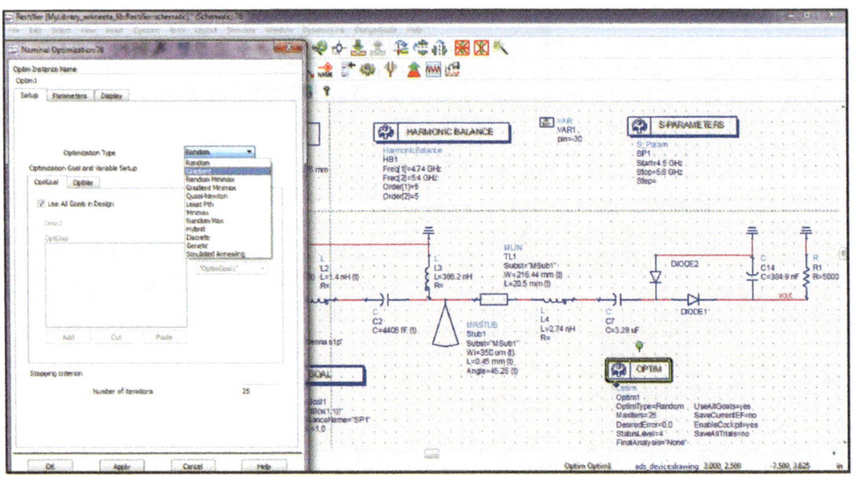

Fig. 5.122 .

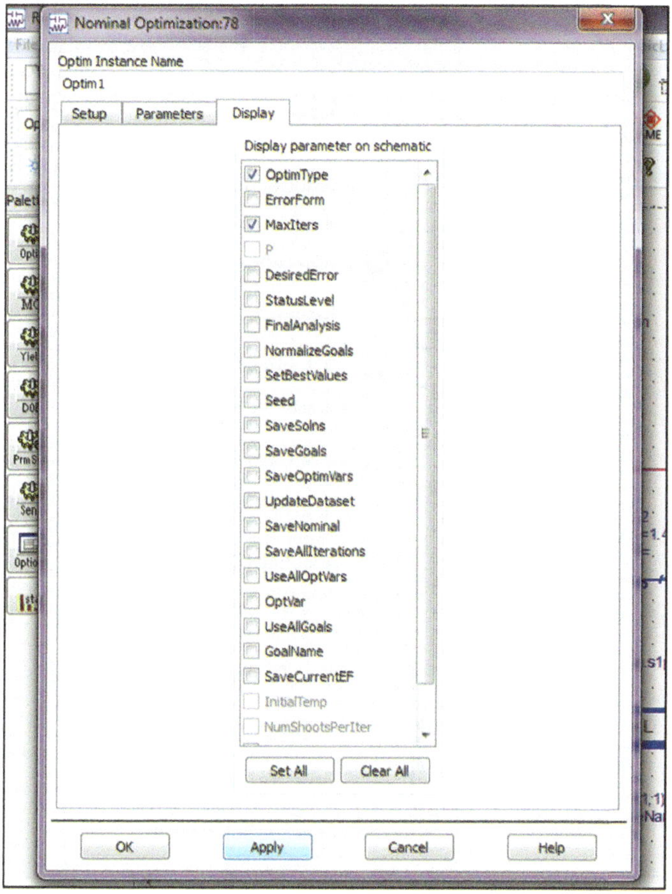
Nominal Optimization:78
Optim Instance Name
Optim1
Setup
Parameters
Display
Display parameter on schematic
OptimType
ErrorForm
MaxIters
P
DesiredError
StatusLevel
FinalAnalysis
NormalizeGoals
SetBestValues
Seed
SaveSolns
SaveGoals
SaveOptimVars
UpdateDataset
SaveNominal
SaveAllIterations
UseAllOptVars
OptVar
UseAllGoals
GoalName
SaveCurrentEF
InitialTemp
NumShootsPerIter
Set All
Clear All
OK
Apply
Cancel
Help

Fig. 5.123 .

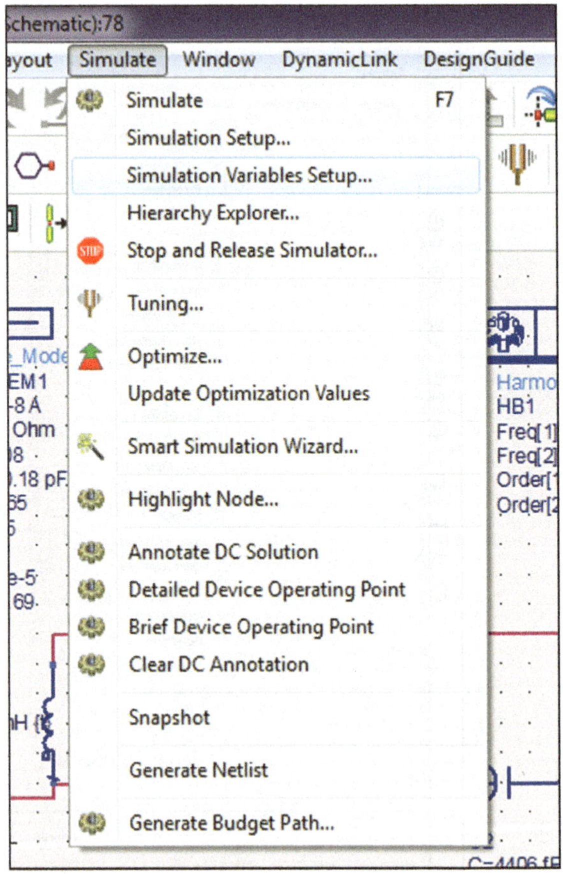

Fig. 5.124 .

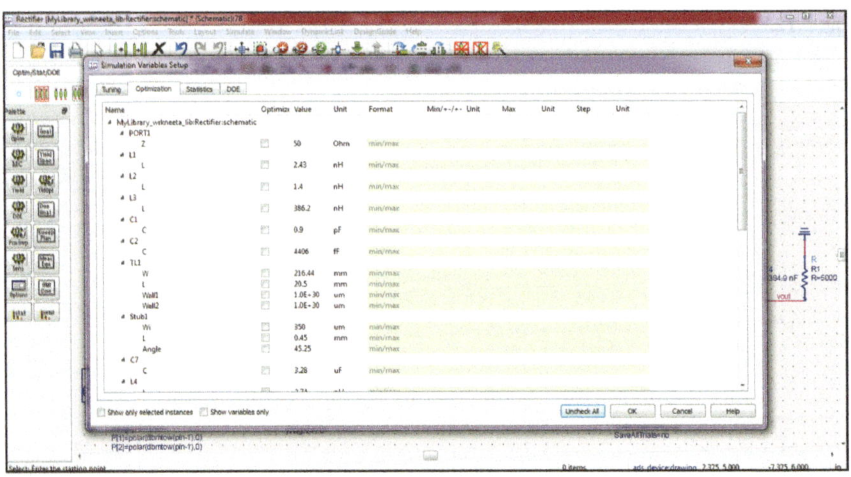

Fig. 5.125 .

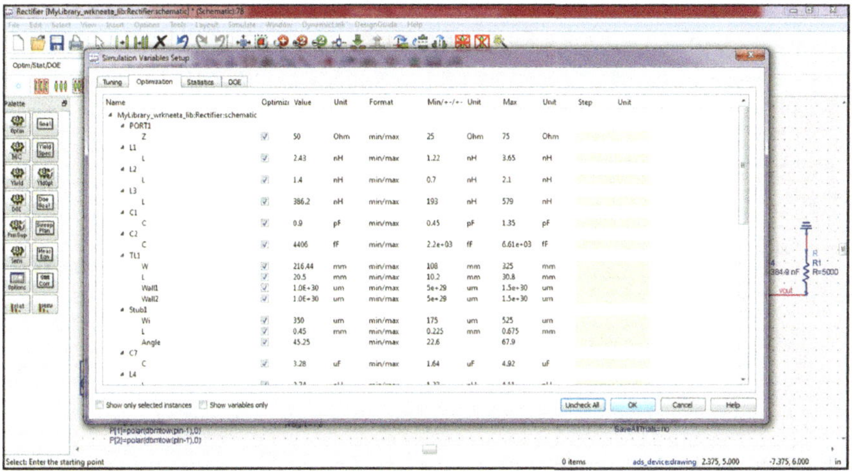

Fig. 5.126 .

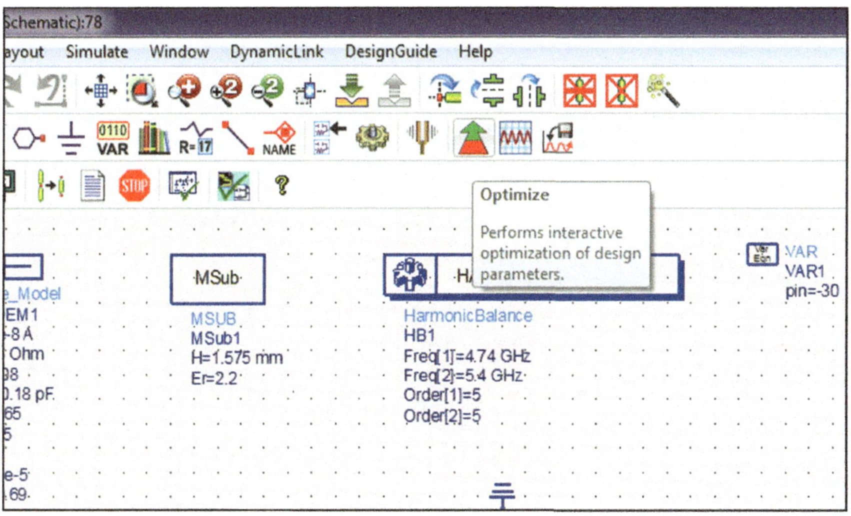

Fig. 5.127 .

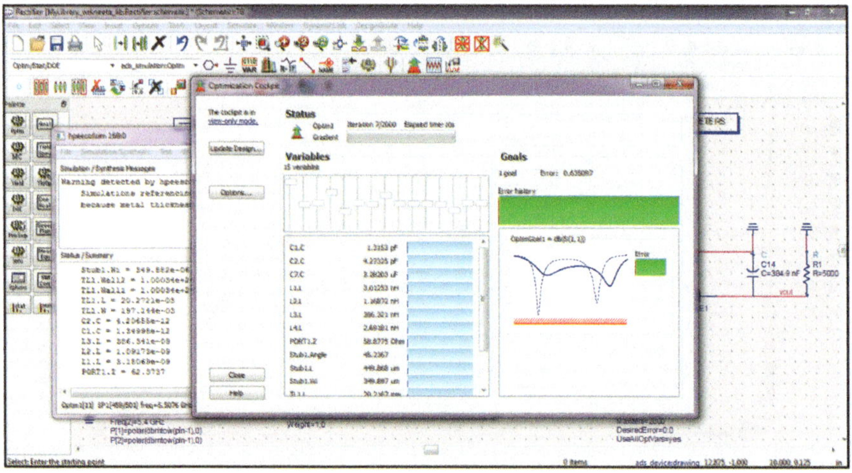

Fig. 5.128 .

The manufacturer's authorised representative in the EU is Springer
Nature Customer Service Centre GmbH, Europaplatz 3, 69115 Heidelberg,
Germany. If you have any concerns regarding our products, please
contact ProductSafety@springernature.com

Printed and bound by CPI Group (UK) Ltd, Croydon, CR0 4YY
30/09/2026
02225330-0001